Don Henry FORD, Jr.

Docket Number P-85-CR-31(2)

Picture Taken April 22, 1985

About the Author

In December 1986, the feds busted DON HENRY FORD, JR., for a second time when he fled to Mexico for his own safety and peace of mind. He served five years in a maximum security federal penitentiary. He now lives with his wife, Leah, in Seguin, Texas, farming, raising racehorses, and writing.

CONTRABANDO

HARPER

NEW YORK LONDON TORONTO SYDNEY

Confessions of a Drug-Smuggling Texas Cowboy

CONTRABANDO

Don Henry FORD, Jr.

With an Introduction by Charles Bowden

HARPER

Author photograph on About the Author page (no copyright) is the mug shot of Don Henry Ford, Jr., after he was first arrested and jailed in Alpine, Texas, April 22, 1985.

Photograph on title page spread: Courtesy of the author.

A hardcover edition of this book was published in 2005 by Cinco Puntos Press.

HarperCollins books may be purchased for educational, business, or sales promotional use. For information please write: Special Markets Department, HarperCollins Publishers, 10 East 53rd Street, New York, NY 10022.

First Harper paperback published 2006.

Library of Congress Cataloging-in-Publication Data

Ford, Don Henry.
Contrabando : confessions of a drug-smuggling Texas cowboy / Don Henry Ford, Jr.
p. cm.
ISBN-10: 0-06-088310-3 (pbk.)
ISBN-13: 978-0-06-088310-2 (pbk.)
1. Ford, Don Henry. 2. Drug couriers—Texas—Biography. 3. Drug traffic—Texas. 4. Drug traffic—Mexican American Border Region. 5. Marijuana industry—Mexican American Border Region. 6. Marijuana abuse—Texas. I. Title.

HV5805.F63A3 2006
363.45092—dc22
[B] 2006041086

06 07 08 09 10 DIX/RRD 10 9 8 7 6 5 4 3 2 1

For my mother who taught me how to speak
& for Leah who restored my faith in women.

ACKNOWLEDGMENTS

Many contributed to my life and the making of this book. It would take another book to mention all of them. For all who have loved me along the way, on both sides of the border, may God's blessing be your reward. You know who you are.

This book would not exist without the advice and encouragement of Charles Bowden. While I will never write with his grace, he is my mentor. The people at Cinco Puntos Press took a jumbled mess—the work of an amateur—and helped me transform it into the book you hold in your hands. I learned a lot from them.

Oscar Cabello is a prince among us. Beto his loyal servant. Photographer Julian Cardona never gets paid enough for his contribution with a camera. Mary Martha and Molly helped when I had nothing to give.

I must also thank my family, one and all. It must be hell putting up with me.

NO ONE EVER GETS OUT OF THEIR OWN LIFE—AN INTRODUCTION

Mars walks along the western mountain before the first whisper of dawn. A cock crows, the burros begin to bray, and then the village dogs bark their cries of territory. Thirty families sleep in the mud buildings just south of the great river, the one Americans call grand and Mexicans call brave, as the man stands in the darkness with that first cup of coffee in his large, rough fist and explains, once again, how he almost had it all and how he lost everything.

"I sensed our days were numbered," he says in his soft, rural Texas voice. "I wanted out. I had dreams of freedom and by freedom, I mean not being told what to do—to live as I want and to work as I want. I wasn't interested in accumulating a lot of things. We were a team here, we tried to break the mold. We were going to do some things."

He's six-foot, the shoulders broad yet rounded by the cords of muscle lacing his neck. He's back after an exile of seventeen years. Now he is forty-six; then, well, he was a young man and he was going to be free. A yellow dog pads by in the darkness, the moon licks the ground with light, the harvest moon that once meant Comanche raids here. On the rocks a half-mile from the house, an ancient painting of the sun, the moon and the stars tells of the time before our time.

"I wanted," he continues, "to fix the ditches in the fields, have a doctor, whatever was needed. Once, a little girl came up and wanted to go to school. But she had to buy the books—there was need all around."

And his business could have done all that and more. Globally, it runs up close to half a trillion dollars a year. Demand never ceases and trade agreements never make a serious dent in the take. The man has had millions and millions pass through his calloused hands, been such a success that various governments took note. But he did this in a business that hardly gets taught in the M.B.A. programs, an international affair where capitalism actually remains cutthroat and the thrills of Wall Street, the Silicon Valley, and the other mirages of some new economy look like the pale copies of life that they are. That said, it is pretty much like any business, and so is the ruin most taste who try it.

He takes a sip of coffee, he leans forward and suddenly his soft voice has gravel in it.

"It was like," he snaps out, "they said, 'you motherfuckers better stay where you are.' I thought the business was a way to break the chains. I didn't want to shoot or kill anyone or have a violent overthrow of the government. I just wanted to steal a little wealth."

And he did. For a spell, his end of the startup was clearing three million dollars a year. And growing. An international airfield opened and brought new options to the village. A network of entrepreneurs reached out to American cities, Europe, Asia, South America. The accountant was some guy in New Zealand. One day a London sea captain retrofitting a ship in Korea bought another ship in Panama, and then suddenly cargo was being offloaded on the Georgia coast. The world became woven together with partnerships and this little village felt the tug of this weave.

Gray light seeps into the village as he talks. Soon the women will be patting tortillas and tossing them onto the comal. Frijoles will start to scent the air, the dogs will patrol their boundaries. The dirt in front of each house will be swept clean and sprinkled with water.

Now he is older, holding that cup of coffee, trying to taste the past and understand how he came to this present. His name is Don Henry Ford, Jr. He lost everything when the business went down, including a decade of his life.

He's the basic American story, the story of success, independence and failure. He had the wife, the children, the normal responsibilities. He refused to be that puppet on a string. Johnny Cash once sang that he walked the line. Well, everyone doesn't. Ford taught the villagers to break their chains. And by that act he all but murdered the village.

* * * * *

A half-mile north of the village, the dead sleep under heaps of rocks in the desert. Four vultures rest on the old tilting crosses. Word spreads of Ford's arrival and the men stream in to talk. They remember things in general and one day in particular, the day a hundred and twenty soldiers hit town and the women and children screamed. For years, Ford has quietly dreaded this moment because he feared they would hate him for what he brought down on their world that day.

But they are all smiles and savor the memory of his time with them.

* * * * *

The sun beats the ground like a drum. Ford eyes the village and drifts back into his life in the business. He remembers the current then flowing through his bones. His eyes say this place feeds me, this place fills me with love. But his head tells him to remember the stress of the business.

"I had a horrible life," he snaps. "I was able to trust no one. I couldn't enjoy things. I couldn't think about having a woman. I had to think about going to jail or getting killed. I only slept with one prostitute in Mexico while I was in the business. I was just afraid. You get involved with someone and then you might say something. So you don't let anyone get close to you. But you just can't go through life alone. It's usually someone you wind up trusting that puts you in jail."

But still, the hungers push him. He reaches like a blind man for what he was and what he wanted: "I just wanted to go to a $100 motel with a $500 whore after a good meal. I didn't want a castle and fancy cars. And then, reality caught up with me. Every whore will spread her legs for you. You can treat your friends, eat what you want. But then you pay with your life."

He knows things they don't teach in business school. He looks up and asks, "Do you know what $200 million weighs in twenty-dollar bills? Ten thousand kilos."

The village sleepwalks past him as he speaks. Children in uniforms moving down dirt lanes to a tiny school, young girls eye boys out of the corner of their hungry eyes.

He suddenly remembers losing twenty-six pounds to a punk in Dallas.

"I was fucking angry," Ford recalls, his anger still acid on his tongue. "I sent out word don't nobody buy from this guy. After about two weeks, he calls and I say, 'Listen, you motherfucker. I know where your mother lives, your family and I'll waste them all if you don't get that shit back here.' And the punk did what I told him to do. But I thought, *What in the fuck has become of me? I'm willing to kill some asshole over this shit.*"

Business is never easy. Rising always takes a toll. Freedom is never free. And the market, well, the market is never predictable. And it is always all about money and then when the money comes it is never enough and when the money comes it don't mean shit. Money is how we explain our actions to others but it never explains them to ourselves.

That is the fact of business.

* * * * *

Make it a story and kill it—a story of drugs, a story of espionage, a story of crime. Or make it money. One man Ford deals with winds up taking in two hundred million dollars a week. His friend in the village reaches out and joins a global combine. Ford himself clears ten million or more. He can't say for sure, he can't recall it all. It comes and it goes.

But it is business. Get in, get out, retire, buy that mansion, have those pleasures. Live free. The same thing, whether a dirt lane in a forgotten village. Or Wall Street.

Money can't buy you love. But it can doom anyone.

* * * * *

Years ago, the government came to the village, built a dam and laid out fields. The campesinos worked like dogs, brought in almost fifteen hundred acres of wheat, threshed the crop by driving trucks back and forth over the sheaves. Then the government took the crop and gave them no money, just a little flour. They gave up. That is the legend of the village. And then, just a few years later, Ford comes and plants a field and the crop is green and soars to the sky. And a plane lands and the cargo goes into a cave and suddenly there is money for meat and beans and maybe even a beer now and then and the women smile and the children skip down the dirt lanes.

And then the army comes, men go to prison, the village begins to die and now only a few families remain. And one day, Ford returns trying to understand how he made it and how he lost it and everyone's face lights up because his face makes them remember when they were men and stood on their own two feet during that brief holiday from their doomed lives.

Ford comes out of the canyon, down the hill, back to the village, past the vultures roosting in the burying ground. And the men say, come back, plant again, it was good then.

Ford's eyes survey the abandoned fields and a gleam appears, just for the flicker of a second, but it appears as brilliant as a marijuana plant racing toward the sun. He was born to make things grow.

* * * * *

Ford wrote this book. He thinks he learned things others should know. He thinks what he has learned is not about drugs, not about prison, not about crime. He thinks he is lucky to be alive and must pay some debt for being alive. So he struggles to get it down in words.

Or he looks for the answers in dirt. He's got fifty horses, dogs, goats, hay to bale. The year he got out of prison he and two Mexicans cut 1,200 cords of wood. He's a fool for work.

He'd like to move back to the village, show the people how to reactivate the fields, raise some food, get free of this world gone mad.

Once upon a time, there was big money, hot women, and a whiff of freedom. After seventeen years, he finally revisits what he lusts for and what he dreads. It is late afternoon now, Beto's woman is making fresh tortillas. Bowls of beans and rice and stew rest on the table in the mud house with three-foot-thick walls. Dogs laze by the door.

Don Henry Ford is a part of a generation that tasted the big money and lost it. He's part of his country's endless lost generations that stumble into a boom and think such moments are reality.

He drives out as the sun sinks. Twenty miles down the dirt track he comes upon a small building in the desert where they sell him gasoline from a barrel. He says he once worked here as a drug smuggler.

The little brown man filling his tank lights up. He remembers those days. Then, when the Colombians were here, he sold 250 cases of beer a week. Now there is nothing.

Don Henry Ford smiles.

No one ever gets out of their own life.

—Charles Bowden

CONTRABANDO

THE BEGINNING

The year is 1980, and I am twenty-three years old. My Suburban floats over the road to Del Rio, Texas. Slightly over two thousand dollars in cash creates a bulge in my billfold. Some of it's mine—some "borrowed" from hay sales at my dad's farm. So—the money is my dad's. Or does it belong to the bank? Who knows? I justify taking the money—after all, this is a business venture to save the farm, isn't it? And I'll pay it back with interest. I rent a cheap motel room and stash an ounce of marijuana and a couple of hundred bucks in the room before crossing the bridge that leads to Ciudad Acuña. I roll three joints. I smoke one before crossing the bridge and take two with me. I avoid the zona de tolerencia—the red-light district—due to previous bad experiences. A bar on the main drag catches my eye and I stop.

The cantina is small—two pool tables and eight or ten tables hosting beer-drinking customers, even at this early hour. I order a beer and begin to play pool, sip my beer and breathe in the smoke of others. Within minutes, a young heavyset Mexican man about my age comes up to the table and challenges me to a game. We talk in Spanish. I've known Spanish from growing up and working on the farms and ranches and from my travels to Ecuador and Columbia with my father. We talk as we play—each of us feeling out the other. I ask if he likes marijuana.

"Sure," he replies.

"Where can I buy some?" I ask.

"I know people. How much would you like?"

"That depends on how much it costs."

"How much it costs depends on how much you want."

"I need ten pounds."

At this, he perks up. "You have money to buy ten pounds?"

"Yes."

"On you? Here?"

"Yes."

"I can get it for you but not until later."

"How much later?"

"This afternoon—today!"

We play a few more games of pool. My game, never very good, is worse than usual. He beats me again and again. I get tired of this and ask if he wants to go out and burn a joint. We smoke about half of the joint, passing it back and forth while driving around town. I put the rest of it alongside the other prerolled cigarette in the ashtray of my Suburban. Smoking the marijuana leaves me comfortable with my newfound friend. He directs me to the zona, insisting all will be well because *he knows people.* We walk into a bar and take a seat. He whispers quietly to several women and a young boy and then tells me he is making arrangements for a meeting with his suppliers.

We drive back to the first cantina. I park, at his direction, in the alley behind the bar. We enter and start playing pool again. By this time, the bar is filling with people, predominately older local types. I buy us another round. My friend glances at my open billfold. We play some more, and he continues to beat me at the pool table.

Two men enter the cantina, dressed in nice clothes and expensive black leather jackets, with slicked back hair and dark sunglasses, looking not unlike the Blues Brothers.

"Here they are," my young associate tells me.

He walks over as they take a seat and speaks briefly with them. He returns and we play another game. I am poised to win. I line up a shot on the eight ball, draw back the pool cue and stroke. The eight ball sails into the prescribed pocket. So does the cue ball.

"Damn!" I exclaim. The two men at the bar watch.

"Let's play another," my friend suggests.

I remove a couple of quarters from my pocket, insert them into the pool table and prepare to release the balls. Then I feel someone grab me from behind.

The son of a bitch!

One man struggles to hold me from behind. I fight—many hours on the end of a shovel and loading heavy bales of hay have left me with a muscled body my fat Mexican assailant isn't able to control. I back across the room and slam him into the wall. He loses his grip and collapses to the floor. I step into the other man, eyes locked on the point of his chin where my fist will connect, like a big cat homing in on its prey. I see fear in his face. He backs up and whips out a badge, holding it up like a shield, or the cross of a frightened priest in the face of an approaching vampire. I stop short of hitting him.

"Police!" he yells, with a hint of panic in his voice.

I raise my arms into the air. The guy I slammed is up again and pissed off. He grabs one arm, his accomplice the other, and they hustle me through the back door and into the alley. They shove me against the wall and try to hit me in the abdomen. I repel them.

"I'll go with you, but not this way!"

"Get in your vehicle," one shouts.

I climb into the front seat of my Suburban. One of them gets behind the wheel. The fat guy rides shotgun. I sit between them.

"Where are you taking me?" I ask, fearing the worst.

"Jail."

"For what?"

The guy riding shotgun opens the ashtray and removes the joint-and-a-half.

That bastard set me up!

We drive to a police station. The building is small and made of adobe but a marked police car sits parked in front. At least they really are police. The alternative could be much worse. I've heard the stories.

They lead me into the building and into a small office containing a desk and two chairs, one behind the desk and one in front. They put me on the one in front. A third man, obviously lower in rank, joins us. The man who had produced the badge in the cantina takes the seat behind the desk and speaks with authority while the other two look on. The fat guy is still pissed off. My captors question me while others outside thoroughly search the Suburban.

The head honcho orders me to strip. As I do, the fat guy searches through my possessions and finds the two thousand dollars in my billfold. I sit down, wearing only underwear and socks. I feel the hundred-dollar bill resting against the sole of my foot.

"Take off your socks and underwear," the man seated behind the desk orders.

I look on in dismay when the fat guy finds the last bill.

"So, you came here looking to buy marijuana?"

"Yes, sir."

"Don't you know that marijuana is illegal in Mexico?"

"Yes, sir. I also know that even the police sell it down here."

"How do you know that?"

"It's common knowledge."

"We throw people in jail for possession of marijuana!"

"Yeah, I'm sure you do. I'm also sure you sell it to others."

At this, the leader motions to his accomplice, the fat one I slammed. He comes over. I look into the beadiest opaque eyes I've ever seen—eyes that show no regard for life. He brings out a gun and points it at my head. I know without a doubt that this man could kill me—and that afterward, he would sleep well.

I don't say a word.

After an awkward minute, the leader speaks again.

"What do you do for a living?"

"I'm a farmer."

"Do you own a farm?"

My mind races. He's trying to see if he can get more money out of me.

"No, but my father does."

"Then why are you down here trying to buy marijuana?"

"Because we're going broke."

"Your father has no money?"

"None. We're about to lose everything we have."

"I'll bet he will send money to get you out of jail."

I lie. "If my father finds out that I'm down here buying marijuana, he himself will lock me up and throw away the key, and then leave me here to rot. He hates marijuana. Please, whatever you do, don't call my father."

The man doubts what I say, I can tell.

"I'm serious. Put me in jail if you have to, but don't dare call my father."

I'm allowed to dress and ordered to sit on the chair again. My captors depart, leaving the third man behind to watch me. This man sits across from me with a holstered revolver on his side. We talk. He seems sympathetic.

The room is terribly hot and stuffy. Sweat pours from my body. Flies annoy us. The guard watching me falls asleep. I weigh my chances. I'm sure I can overpower him and get his gun. I can't hear anyone, and I know my Suburban is right outside the door.

But, are the keys in it? Is the guard's weapon loaded, or is this a setup? Do I want to risk being shot or shooting someone else to get away?

I sit. And sweat. And worry.

The men return after several hours.

"If you plead guilty to possession of marijuana, I have been given authority by the judge to let you go with a fine. Otherwise, you go to jail. I suggest you take this offer. It will be much more difficult to fix this problem once you go to jail."

"OK, I plead guilty."

He produces a piece of notebook paper with a written confession. I sign it.

He returns my possessions, counts the money in my presence and hands it back to me.

"The fine will be two thousand dollars."

Son of a bitch! I think as I count out 20 one-hundred-dollar bills—nearly all the money I have in the world—and hand it back to him.

"God is watching," I tell the man.

He looks curiously at me, then hands me the keys to the Suburban and tells me to follow him to the bridge, ordering me to go straight home without telling anyone about our arrangement. I drive through Ciudad Acuña behind a Chevy Blazer. He turns off at the bridge. I continue north. Crossing the bridge, I think about getting a gun and returning to kill every one of the bastards.

Others might have taken such an event as a sign to quit. I, on the other hand, thought of ways to get the money back. All the way home, the only thing I could come up with was to try again.

Back home at my father's farm, I was glad to be alive and free, but still determined to be a smuggler. I didn't tell anyone what had happened. I was ashamed to have been hoodwinked, and I sure as hell didn't want to explain to my wife, Cheryl, how I managed to blow two thousand dollars we didn't have.

* * * * *

We've all heard the stories of the man who goes down to the crossroads to make his deal with the devil. The decision I made at this crossroads—to smuggle marijuana out of Mexico and into the United States—put me on a path from which it seemed like there was no escape. Not many who take this journey survive to tell the story—and those who do are forever marked by the experience.

I saw many things during those few years of my life—not all were bad. The business provided large sums of money when things went right—the cash opened doors to the finest hotels and restaurants and the legs of the prettiest whores. That same money provided me and my friends the best drugs available. I traveled to parts of the world that I would never have otherwise known. While I was there, I saw their beauty—tasted, felt and smelled their essence—and met both good and bad people. To say I didn't have fun would be to lie.

I bought and sold many tons of marijuana and smoked enough of it to wear out several sets of lungs. I felt the rush of adrenaline as I crossed borders heedless of the cops assigned to guard them. Pride and contempt for law enforcement officials grew in me as I beat them time and again at the game—in all perhaps a hundred times. I walked around with thousands of dollars in my pockets and the ability to buy whatever I wanted. I looked down haughtily on those with "real" jobs—slaves of the system—who struggled endlessly without hope of breaking free. Their payday was my payday—they rushed to banks with their paychecks and then found me to buy a little relief from their misery with cash in hand. The day would come when I weighed the money to count it.

But the sweetness of success was tempered by bitter failures. To love those who betray that love, like the whore who wants only your money, or to trust those who want only the high of your drug or the wealth to be made from selling it, inflicts deep injuries never to be completely healed. Not in this life.

For me the days came when the cops won, when rival smugglers and dealers beat me at the game and the whore I loved found another. I tasted defeat time and again and discovered the horrors of being locked in a prison—knowing my family struggled to survive without a husband, father, son or brother. I watched friends die needlessly. I came very near dying myself. The day also came when the drugs I craved no longer brought relief or a high but only the senseless need of an addict and the damage to mind, body and spirit for which they are rightfully notorious. And the money was gone.

I witnessed the front line of this War on Drugs we fight—a place where the difference between the good guys and the bad guys ceases to exist, where all who enter are victims, a place where nothing heard can be believed, nothing seen can be trusted, and a place where the only goal becomes survival.

A place where no one wins the war.

★ ★ ★ ★ ★

The following is an account of events as I lived them. Not all of the story is here. It took years for this stuff to happen and many boring days occupied the space between events. I omitted some episodes due to possible repercussions that might come to friends or me to this day. I left out other things because I knew it would be hard for anyone to believe I didn't quit sooner—especially if I described all the times I got arrested in Mexico.

But this story isn't only about drugs or me. Not entirely. It can't be. It's about a world gone mad—it's about fire and smoke and sweat, blood and dirt and blisters, empty stomachs, sick children, the feel of wood, the smell of a horse, barbecue, grains, and fruit—and smooth brown skin and glistening black skin and white skin burned red, and sun and freezing cold and water, and spirits and plants and sky, stars in the night, and love. We have forgotten where we came from. I have to remind myself. I can't forget. We must not forget.

But we do.

And for some reason, we look for our answers in drugs.

Parts of my past are very hard to tell. Looking back, it's easy to see that I became a predator. I abused women. They abused me. I took advantage of the weak—those addicted to drugs. My children suffered because of my actions. I'm not proud of this. But this isn't a novel. I have to tell the story that was—not what I would have liked it to have been. I was no hero.

And I don't want to glorify what I did—only to tell the story as it happened, to the best of my ability.

CHILDHOOD

I was born in February 1957 at a hospital in Midland, Texas—not four blocks away from the childhood home of George W. Bush. West Texas was in the middle of a drought, and huge sandstorms filled the winter sky. My dad was twenty-seven, my mom twenty-two. I was their first child. My grandmother Ford took one look at me and nicknamed me "Cowboy." This proclamation proved prophetic.

My dad was born in 1929, right before the crash of the stock market. The Depression forged his way of thinking, like it did for so many of his generation. His father had always been involved in farming. My grandfather migrated from Kansas to Texas, where he kept up that way of life. My dad spent many grueling hours working on farms and ranches under his father's watchful eye, earning almost no money. Dad became convinced this was no way to live. An education seemed the way out of that life. But an education cost money—money he didn't have. He enlisted in the army and tested high on an IQ exam. The army put him through combat engineering school and an intense language course, which transformed his crude border Spanish into a more sophisticated form. He then served in the Counterintelligence Corps in Panama during the Korean conflict. After he got out of the army, he enrolled at the University of Texas under the G.I.

Bill and studied geology. There he met my mother, a fellow student and an English major, and they were married. I've always suspected he maintained contacts within the intelligence community.

Finding a job as a geologist proved difficult. My dad moved us to Farmington, New Mexico, before my second birthday in order to find work. By the time I was five, his pursuit of oil led to Roswell, New Mexico, where I got my first taste of that awful thing they call school. Outside of getting to see friends, P.E., recess and a few girls who captured my interest, I found little I liked there. Nonetheless, learning came easy to me and I made decent grades without doing much in the way of homework.

By the seventh grade, Dad moved us again, this time back to Midland, where he took a job with a corporation out of California by the name of Cayman.

The only constant during these years were the trips to Abilene to visit my grandparents. My grandfather Ford played a profound part in the person I wanted to be. On occasion he took me and my brother Bill to a ranch he owned between Tuscola and Ovalo, Texas. We got to hunt, fish, ride horses and work the soil. My granddad also raised a garden at his home in Abilene and put me to work plowing, making rows, planting and harvesting vegetables. He showed me how to graft pecans and take cuttings off of plants and root them in soil—all the while presenting it to me as work he needed done, work I got paid for. I mowed grass, hoed weeds and sprayed for insects. We harvested peaches and pecans, eating them in huge quantities. We ate watermelon in the evenings in his yard and took a trip almost every day to a local establishment that served root beer floats at curbside.

My granddad could not let a good-looking set of legs on a woman pass by without comment, so he didn't try. I was always embarrassed for the girls. But the way he pointed out their attributes was offered with a smile and was entirely genuine, and most of them appreciated the compliment.

In the middle of the eighth grade, my dad moved our family to Quito, Ecuador, where we spent close to two years. I was snatched out of one culture and thrust into another. I learned there are people in this world who don't enjoy the things we take for granted in the United States. While my dad operated mostly in circles of the rich and powerful, I got to know the Indians and the poor on a personal level, and I vowed never to forget them.

I remember helping native workers while visiting a jungle base camp, loading supplies destined for oil rigs onto helicopters. They were uncomfortable with my help, but I insisted. Afterward I sat among a group of maybe twenty of these men, describing my world and my life while they described theirs. One of the men declared, "You'll come back some day to help us."

Every day a group of children came by our house to beg for food, and my mom gave them a liter of milk and a loaf of bread. Their eyes lit up. The youngest—over two years old—couldn't walk. Within a month after getting this daily ration, he walked. Indian women carrying babies on their backs pushed heavy wheelbarrows of cement up steep wooden ramps while I sat and stared out the window of the missionary school I attended. Some parents disfigured their children and sent them out to beg on the streets. Visions of their innocent eyes and tiny grimy hands still haunt me. Our family's maid worked for the equivalent of twenty dollars every two weeks and raised a family on it—her husband had gone to the United States and abandoned her. This was above-average pay for an Ecuadorian maid.

Indians guided my dad high into the Andes in search of gold—14,000 feet and more. My brother and I tagged along. They walked while we rode burros. The Indians took us into their homes, shared their food with us and protected us from our ignorance in potentially hostile country. Years later, I would find similar types in the mountains of Mexico.

Later, back in Quito, I watched as my dad answered the phone. A look of shock, pain and horror crossed his face as he received the bad news: one of the wells being drilled had blown out and exploded, killing the derrick hand in spite of the heroic efforts of a helicopter pilot to get him off of the rig. "What about the geronimo?" he asked.*

No one had bothered to install one. I'm sure my father felt responsible.

We returned to live in Conroe, Texas, but in two short years I had fallen out of step with my own society and my peers. I never quite caught back up. I never really wanted to. My dad became wealthy, but our wealth shamed me. I wore boots, chewed tobacco and rode horses almost every day. I began to compete in rodeos and made spending money by hauling hay for local farmers. I didn't want to belong to the world my spoiled rich schoolmates belonged to—a world of golf and tennis, water-skiing at the lake behind expensive power boats, social events like proms and wearing just the right brand of clothing. I rarely drank alcohol, which kept me from being accepted by the rednecks and working-class kids—for that was virtually all they wanted to do.

---

*A geronimo is an escape mechanism (a pulley with a handle) that is attached to a cable which is connected to the rig near the derrick at one end and the ground at the other. It can be used to bail off of the rig by the derrick hand in the event of a blowout, like rappelling off a mountain.

In the summer between my sophomore and junior year in high school, I moved to Baird, Texas, to live with a Methodist minister my dad hired to take care of a cattle ranch he bought. The preacher needed help building fences, working cattle and other assorted chores. I enjoyed that life. I spent that summer and the next working for my dad and the preacher, and I decided that raising cattle would be my profession.

After graduation from high school, my friend Tim Peterson and I were given the opportunity to go to Colombia to work for a summer for Farmland Industries International. They had bought Cayman's properties and retained my dad. He arranged jobs for us but—just about the time we arrived—a dispute with the government shut down all the oil companies, so we hung out in the Putamayo basin waiting for work to resume. We spent our time among poor Indian people, learning how they lived. Farmland Industries treated workers far better than average—at my dad's insistence they built schools, funded orphanages and tried to help usher these primitive people into the modern world. Nevertheless, long-standing traditions in South America—the huge gap between the haves and the have-nots, and the indifference the rich have for indigenous peoples—remained intact. Those who tried to change this encountered stiff resistance.

If an American company paid workers more money, the workers of rival companies—often owned by those who ran the government—would want more as well. There were limits to what they could pay. However, if a company failed to pay enough, they might find themselves the victim of antigovernment revolutionary forces. *Enough* was never enough to change the existing social structure. The rich got richer and the poor stayed poor.

In the end, the bottom line for American companies was making money—lots of money—with little regard for the environment. Trees were slashed, roads cut. I watched while thousands of gallons of oil were sprayed onto nearby swamps and marshes as a well was being tested. The contract drilling company's men whooped and hollered as the oil blasted out of a large pipe with such intense pressure that it traveled half the length of a football field before hitting the marshy ground. I asked why they wasted the oil and didn't get a satisfactory answer. Oil companies spread sticky crude on dirt roads daily to help settle the dust. Some of that oil had to end up in waterways. This practice is illegal in the United States.* For good reason.

Tim and I returned to the United States and my father's ranch in Baird and

* Texaco is currently being sued by groups in Ecuador and vehemently denies ever having done many of these things. I've seen the evidence and I am here to tell you they did.

then something bad happened. I won the bareback bronc riding competition at the Baird Junior Rodeo.

* * * * *

As the national anthem plays, I close my eyes and say a prayer. Tim pulls the cinch to my bareback rigging while I sit on the back of the nervous bronc. I've drawn a blue roan stallion known to be a high bucker. He likes to spin and jump without covering a lot of country. I hold the rigging in place with my gloved hand. Tim ties off the latigo while I work my hand in and out of the rigging a few times to heat the resin.

The announcer calls my name and the blood and adrenaline kick in. All fear evaporates. I feel alive—on the edge, where every second counts and is precious. The men who work for the rodeo company approach—one from behind to pull the bucking cinch wrapped around the roan's flank and others in front to open the gate and swing it aside. They want the horse to throw me. The horse senses the time is near and moves back and forth, raising his head as if he wants to jump out of the chute. He's done this before.

I jam my hand into the bareback rigging as far as it will go and then twist it back, squeezing the handle. I clamp my hat down, slide forward, nod my head and yell, "Let's dance!"

The chute opens wide. I thrust my feet forward and plant spurs into the horse's neck. He takes a short hop out of the chute and then explodes, high and hard. I feel the power and hear his fart and his grunt as he gives me all he has. He returns to the ground and my feet find his neck again. He makes another short hop and then explodes skyward again. My feet fly backward, but my ass remains planted right behind the rigging. I watch the curve of his neck and the hair of his mane. This time when he hits the ground, he turns sharply to one side, trying to lose me, but I stick with him. I know I have him—and he does too.

* * * * *

I missed the award presentation four days later because I was in the hospital having blood-filled urine drained from my bladder. Tim Peterson received the trophy buckle on my behalf.

It took me several years to discover that I was not cut out to be a professional bareback bronc rider, but I tried damn hard to be one.

* * * * *

I entered Texas A&M University in the fall of 1975 to study animal science and to rodeo. I also signed up for the Corps of Cadets. This was not something I wanted to do, but my dad insisted. He was paying for my education so he won the argument. I was late getting registered and had to take what classes were still available—courses like Old English Literature. Like somehow Old English literature was going to help me raise cows.

While my dad was responsible for me joining the Corps of Cadets, I came up with the brilliant idea that the Marines would be better than the Army. A barber shaved my head. I was fitted for a uniform and then reported to the fourth floor of a dorm—home of the N-2 Neanderthals. I noticed a group of upperclassmen with sly grins checking me out on the way up the stairs. That evening two sophomores came to my room and handed me a manual. We were told that the next day we would be required to know everything in it and to abide by all the rules—which, of course, is impossible by design.

Overnight we had to walk differently, talk differently, dress differently. We even had to shit differently. We were required to ask permission to flush the toilet, which was no longer a toilet but "the head." Our beds had to be made to exacting specifications. We ate in a chow hall seated at attention and were not allowed to look around. The upperclassmen referred to freshmen as fish. Suddenly I was "Fish Ford."

The first night they woke us and forced us to run a gauntlet through the parade ground. We yelled and screamed as we ran while they struck us with towels and belts. We were deprived of sleep and made to do torture exercises—regular exercises extended in time to dangerous levels—every time we broke a rule. Even if we didn't break a rule, we did torture exercises because some freshman always made a mistake, and we all had to share his punishment.

This system is designed to break down a recruit both physically and mentally and then rebuild him in the mold of a soldier, only I didn't do well with the first part of the equation. While others cried in their beds at night, I plotted revenge.

A personality conflict developed between my squad leader and me. He'd kick my boots and smudge my brass during inspections to make sure that I would have to go on forced runs each weekend. No matter how hard I tried, I failed inspection.

I chewed tobacco and saved the spit in a brass spittoon, allowing green slimy mold to grow on it. One night I snuck into his room and doused him in the face with this concoction. By the time he woke up, I was gone, and he couldn't prove that I was the one. But he knew. And he intensified the abuse. Each night he had me in his room doing torture exercises.

Another friend of mine, a former Golden Gloves boxing champion from Cleburne, Texas, got similar treatment. I watched as he snapped one morning and knocked out our squad leader with a vicious punch. I made the mistake of laughing about this. My friend got kicked out of the Corps. If I had to choose one man from that company to watch my back in a fight, it would have been that redhead from Cleburne.

One day they made the mistake of letting us play football—freshmen vs. sophomores. I spent the entire game knocking my squad leader on his ass, which only served to make him hate me worse. He intensified the harassment.

All of this began to take its toll. At mid-term I still had three Bs and one A. But I was exhausted and began to fall asleep in class. I was property of the Corps all day—except when I was in class—and most of that time I was in trouble.

Texas A&M had a tradition of freshmen cadets trying to kidnap the CO of their company. It was the job of upper classmen to prevent this from happening—a war game of sorts. The tradition had been discontinued—we were told that such a thing wouldn't be allowed to happen. And, if it did, the entire company would be disciplined.

But I was already in trouble all the time anyway. A group of guys and I began to hatch a plot. Right before we were about to execute our plan, upper-classmen arrived at my room, one by one—sophomores, juniors and seniors. Someone had told on us. Each of them warned me that they all knew what we were up to and that this absolutely would not be tolerated. *We're talking jail time.* They left convinced I was dissuaded.

Around midnight every night our CO showered after everyone else was fast asleep. That night he started down the hall in a towel and then spotted two freshmen waiting to jump him. He turned, ran back to his room and slammed the heavy metal door. But we had a backup plan. The guys dumped an entire bottle of ammonia under the door.

Our CO tied sheets together and rappelled out of the window of his fourth-story room, clothed only in shorts and combat boots. We were waiting just around the corner at ground level. We jumped him, forcefully restrained him and then pulled a cotton sack over his body. Then we threw him into the bed of my pickup. I drove through the sacred grass of the courtyard—which we weren't even allowed to walk on—cutting ruts. My CO bounced around the bed of the truck like a sacked pig.

The rest of the upperclassmen woke up just in time to see us speed away.

It was cold that night, about 33 degrees. Freezing rain fell on our poor CO all the way to Freeport some three hours away. By the time we turned him loose, he was shivering and almost purple. We released him just like we had taken him—in combat boots and shorts. Then we returned to Texas A&M to await our fate.

My days as Fish Ford were numbered. I remember the look of fear on the face of my squad leader as I walked back to the dorm. We had beaten them all, and they knew it. And they were in trouble too.

Our CO got back the next day. He was angry but I think he was also amused. Somehow word of what had happened never made it to the staff at A&M.

I had thought that what we went through as freshmen before this incident was tough. But that was just because I hadn't seen what lay in store for us afterward. Everyone who participated was required to sleep in a single room each night—twelve of us on a hard tile floor in jockey straps and combat boots with the heat turned up to about 90 degrees and the lights on. We did torture exercises in the shower room with all the spigots running pure hot water to create steam. Any time we went out of our room, we had to crab-crawl or duck-walk. An associate had to hold our hand as we took a shit. And on and on. . . .

I skipped all my classes so I could sleep during the day. When you skip classes at Texas A&M, you flunk. At the end of the semester I had the first F I had ever made. In fact, I had all Fs.

I had reached a crossroads and made a sharp left turn. The next crossroads was not far ahead. That's the way it is for an eighteen-year-old in this world—one critical decision after another in rapid-fire sequence. And there's no turning back.

INNOCENCE LOST

I was severely reprimanded by my parents. I transferred to Cisco Jr. College and began to make good grades for a time—that is, until I met Cheryl. I was living and working on my dad's ranch in Baird, Texas, between sessions in school and rodeoing on the side when she came along. Cheryl was pretty, petite, brunette and one of the first women to notice me. She was the mother of a young boy and she smoked marijuana. I had never tried marijuana—in fact, only a few months before I had threatened to beat my little brothers when I learned they had been smoking dope.

Eventually I succumbed to temptation. I got my first taste of a real live pussy and marijuana from her, all about the same time, and soon found myself married to both in spite of the fact that everyone told me I was making a mistake. Somehow the marijuana made the pussy better. After I began using it, I found a lot of things it seemed to make better, things like getting out of bed each day. I lost interest in school and dropped out.

I took Cheryl's son Dion as my own and our family was formed. She got pregnant right after we were married. We moved into my father's ranch house in Baird, Texas.

My first job as a married man was growing produce. I made up five acres of rows with nothing more than a garden hoe, planted vegetables and damned near starved to death. I remember going to grocery stores with a pickup load of squash, black-eyed peas and cucumbers and coming out with one sack of groceries and enough money to buy gas to get back the next day.* And then being unable to straighten up at the end of a row after picking peas but turning around to begin the next row anyway. I remember hot sun, sweat, blisters, insects and the wonderful smell of water hitting tilled earth and the aroma plants make when they receive that water and stretch their leaves toward the sun. It's a shame to love something that will ensure you are poor for the rest of your life. But that's exactly what I did. I learned to love raising plants and animals.

Since I couldn't make enough money to support my family growing vegetables, I took the seeds from a bag of weed and raised a small indoor crop of marijuana. This made money—not a lot, but I saw the potential. We moved to Abilene. I insulated buildings, then worked on a seismograph crew and roofed houses. I got a job as a swamper moving oilfield rigs and then tried being a roughneck. I supplemented meager earnings by growing and selling marijuana to others who shared my affinity for the stuff. No matter what job I had or where it took me, marijuana became a part of what I did.

Cheryl was a knowing and willing participant in all of this—in fact I bought my first bag of marijuana from her brother. Once I had some to sell, I sold it through him as well. But my smoking habit became worse than hers.

The day came when my son Dusty was born. Our children were being neglected. I disliked the various jobs I had found.

I determined that Oregon was the best place in the United States to grow marijuana and was also good cattle country, so we struck out for the Northwest. I financed our move with a grubstake—I harvested the eighteen ceiling-high marijuana plants that I was growing in our walk-in closet. Sort of a homemade hydroponic operation.

I hoped that we could find a place to grow marijuana.

My first job in Oregon was on a huge cattle ranch where I earned a whopping $525 a month for six twelve-hour days of hard labor a week. One day, Don Martin—the owner of the ranch where I worked—sent me with a neighboring rancher, James Robertson,† and a hand of his to repair an irrigation ditch that

* I received between six and eight cents a pound for my produce. The store sold them for thirty-nine cents a pound.

† Not his real name. James Robertson was a friend and mentor. I don't know if he's in jail, alive or dead. Wherever he is, I wish him well.

supplied water to our fields and to his. I took one look at Robertson and knew he smoked marijuana. Both James and Patrick, his ranch hand, sported long beards and hair and looked like hippies.

As soon as I got into their truck, I said, "Well, it's about time I found someone who smokes pot. I haven't had any for a week."

I think James was taken back. He replied, "Yeah, we smoke dope, but what we're really into is Jesus."

Jesus was all right with me too.

We drove high into the mountains, talking all the way. After we finished repairing the ditch and removing all the trash that had accumulated over the winter, Patrick produced a film canister of potent Colombian Gold marijuana and rolled a joint. He then gave me the remainder of what was in the canister to take home.

This proved to be the beginning of the end of my job with Don Martin. I spent more and more of my free time at Robertson's ranch getting high. Eventually he offered me a job and I accepted. To look at James Robertson, you'd think he had very little money. He wore what he called dungarees and what I'd call blue jeans. He drove a battered old pickup. I learned that both he and his wife were originally from Florida. And that he had amassed a lot of money unloading marijuana from Jamaican and Colombian ships off of the coast of Florida. Robertson's dad had been a fisherman, so James knew all the nooks and crannies of the inland waterways. He and his high school friends, all of them sons of fishermen, would go out with a fleet of small fast boats and unload the ships and then take the merchandise to waiting storage facilities and vehicles. He had dropped out of school at seventeen and was a millionaire by age eighteen.

Eventually he became one of the largest marijuana brokers in the United States. Feeling the heat, he decided that it might be a good idea to move a little farther from the action, so he moved to Oregon. By the time I met him, he rarely handled the products he sold. He worked full time as a farmer and a rancher. A couple of times a year he'd disappear for a few days to receive a major shipment—some, I was told, in the sixty-ton range. Sometimes the merchandise was Colombian. Sometimes it was Thai stick. Or Lebanese hashish. Whatever it was, it was good—he dealt only in high-quality dope. Others did most of the work.

James did not deal in hard drugs—just "medicinal" herbs. In Jamaica he had studied the Rastafarian religion. I don't think he believed all of it, but he found the pot-smoking part—considered a sacrament by those who practice that faith—to his liking. I did too. Fit right in with my plan—staying high at every waking moment.

James wanted to grow marijuana. While he called himself a rancher, he had missed out on the education a person gets from being raised on a ranch or a farm. That's where I came in. I helped him raise cattle, cut hay, train horses and grow marijuana. He was a gifted mechanic and a hard worker. He taught me how to run a chain saw and to fell timber and how to use draft horses to drag logs from the forests. We slaughtered our own meat, milked our own cows and made butter and cheese. We grew as much of our own food as we could and preserved it by drying and canning. We raised grains and ground our own flour, baked bread in wood stoves. We tanned deer hides and made buckskin clothing. I lived in a teepee. All of this to prepare for the "end times" that were sure to come any day, just like Jesus said. We ate no white sugar or preservatives—didn't want to contaminate our bodies. But we smoked all the marijuana our lungs could stand—and then a little more.

And—on occasion—I drove around the country in vehicles loaded down with tons of dope for James, helping others to stay high as well. We couldn't quite wean ourselves from money.

Cheryl got pregnant again. We returned to Texas and Joshua was born in our apartment. We didn't go to the hospital, but instead used a midwife—mostly to save money but also because we thought it was a good thing. Two weeks after he was born, we returned to Oregon.

James and Cheryl didn't get along. So I rented a place and worked for James on the side. Neither of us was happy with the arrangement.

Eventually the call of West Texas became more than I could bear and we returned home. I decided to quit smoking pot and to go to work for my dad on a West Texas ranch he had bought near the tiny town of Bakersfield, Texas.

THE FARM

Our Bakersfield farm was located in an area known as the Trans-Pecos region of Texas. The land was flat, with rich, silty loam soil. It was surrounded by flat-top mesas and looked somewhat like the former floor of an ocean. A closer look at the rocks on those flattop mesas revealed evidence of sea creatures and lines that had at one time been boundaries between land and sea. Now the only water to be found in the Bakersfield Valley ran underground. Minneapolis-Moline engines, which ran on natural gas also taken from the ground nearby, sucked this precious substance through wells to the surface. Without this water, the ground above—although rich in nutrients—would support little growth of any kind. With it, incredible yields were obtainable. But incredible yields didn't necessarily mean a profit—in fact huge yields just meant you lost less money. Everyone except a few diehards had given up farming the Bakersfield Valley.

The Pecos River formed the northern border of our farm, which officially meant we were west of the Pecos, where in times past it was legal to kill Mexicans without getting charged with murder. Old-timers still tell stories of people who dragged a body across that river to avoid prosecution. In their day, the Pecos was said to have been a mighty river, but upstream dams put an end to that. Now it's a pitiful stream of extremely salty water moving through a thin strip of salt cedar trees—almost the only kind of tree that will tolerate the levels of salinity found there.

During the fifties, farmers cleared the land of the sparse vegetation—mostly brushy mesquite and cacti—and drilled into the underground streams. They used sour natural gas from nearby oil wells to run irrigation engines. It was free because there was no processing plant nearby to clean the H_2S from the gas, and it couldn't be sold in its poisonous state. For a time, the farmers did well in spite of the harsh growing conditions. They primarily grew cotton, which was stripped by the hands of thousands of Mexican migrant workers.

"Reforms" to immigration laws put an end to that—the *bracero* program was eliminated and Mexican workers were no longer allowed to come and work legally in the United States.

The area was notorious for rough weather—wind, lightning, tornados and hail usually accompanied the few rains that came each year. Crop losses were common, but "picker cotton" ripened in various stages, and the farmers could remove bolls from the green and still-growing plant, which is extremely sturdy. Several pickings a year were done. Modern times brought harvesting equipment, which took the place of handpickers. These mechanical "strippers" picked the cotton from dead or dying, fully ripened plants in order to perform effectively and to keep from staining and ruining the cotton. The newer "stripper" varieties of cotton were bred to ripen and die before harvest, meaning the plant was in a fragile, vulnerable state for some time before it could be harvested.

Chances of getting through that vulnerable state in the Bakersfield Valley were slim, particularly if you relied on contract strippers to harvest your crop.

* * * * *

We had good hired help on our farm. Especially two permanent hired hands.

One was José Chávez, a knowledgeable farm worker and hardworking cowboy. A native of Chihuahua, José was an angular man with a reddish complexion and hazel-green eyes. He had once been convicted for smuggling marijuana and was still on parole, but he was certainly less than repentful for this transgression. While José didn't smoke marijuana himself, he had sons who did. My promise to quit smoking marijuana lasted all of two weeks.

Inés Pérez was heavyset, dark-complected, with big brown eyes that radiated light from within and the large calloused hands of a man who has worked his entire life. He was about forty, born and raised in the United States, the son of Spanish-speaking Mexicans.

When it came time for Inés to go to school, he spoke no English. He went home after the second day and never returned, going to work instead in the fields at the age of six. By forty, Inés was still unable to speak English but was proficient at operating all kinds of farm machinery. Inés could take apart and put back together virtually any piece of mechanical equipment on our farm—an area in which I was sadly lacking. He was a loyal and devoted hand, but worthless for a day or two each time he got paid.

During the week, Inés limited his drinking to after hours but on the Saturday afternoons when he got paid, he'd set out on a thirty-six hour non-stop binge—always Coors Light. He claimed any other brand would make him sick. On Monday, he came back to work with a hangover—or still drunk—and toughed his way through the day.

Inés loved me in his own way. Although I felt slightly out of place, I was welcome to attend his weekend gatherings. He always cooked out—fajitas were his specialty. During this time, fajitas were not so popular and could be bought cheap. Fajitas are made from a piece of meat known as a skirt steak—a thin, membranous piece of lean meat removed from the inside rib cage of a beef. They were a less-desired cut and were usually made into hamburger. If cooked like a white man normally cooks a piece of meat, the resulting product is so tough it can scarcely be eaten. Mexicans got around this by marinating fajitas in acidic sauces which contained lime juice and/or vinegar and a splash of oil. Some people added fruit juices like papaya which contain enzymes that break down the tougher parts of the meat, leaving it tender. Each cook tends to put in certain *secret* ingredients. Inés's was beer. I like a little soy sauce myself. After a long soak in the solution, the fajitas are grilled over mesquite coals and sliced thin, against the grain.

The resulting taste is wonderful.

The dish is traditionally served with long green chilies roasted over mesquite coals and peeled. It is always accompanied by plenty of fresh homemade flour tortillas. Often onions are grilled over the fire and occasionally tomatoes as well—just long enough to remove the skin. When chilies are not in season, *pico de gallo*—made from diced serrano or jalapeño peppers, chopped onion, tomatoes and cilantro—suffices.

Inés's cure for a hangover, usually served on a Sunday, was *menudo.* Many West Texans love this dish, but I never quite took a liking to it. *Menudo* is made from tripe—a nice way of saying intestines—which have been thoroughly cleaned and boiled, and are cooked along with hominy in a spicy red broth. I don't really know all the ingredients that go into *menudo,* but think it merits mention because it is a part of the culture and the life of a man like Inés who swears wholeheartedly by its effectiveness to prevent and cure hangovers. Inés had quite a lot of practice at curing hangovers. I found it preferable to not get drunk in the first place and diligently avoided eating *menudo* after trying it one time.

★★★★★

Inés's wife, Rosa, claimed to be a descendant of Cherokee Indians. She was a small, dark-skinned woman with big teeth, long shiny black hair and prominent cheekbones. For lack of a better word, I would describe her as scrappy. Cherokee Indians were known to be tough folks, and some of that toughness remains in their descendants. They also subscribe to many beliefs members of the white race find unusual.

After Lisé, my oldest daughter, was born, she cried from teething pain. Rosa brought a necklace made from rattlesnake bones, insisting I should place this around her neck to ease the pain. Lisé eventually got over the pain, but I doubt the bones had anything to do with it. Then again, who knows?

Rosa's specialty was enchiladas, but hers were not like what most people think of when they think of enchiladas. She made them with thick homemade corn tortillas, stacked one on top of the other like pancakes, covered with an extremely hot, thick, deep-red chili sauce made from reconstituted long dried red chilies. These chilies are often seen in *ristras* decorating the walls of Mexican restaurants in the Southwest.* Rosa's enchiladas were topped with fresh white cheese—never meat. Some people adorn this stack of food with a fried egg, cooked sunny side up. I also like chopped onion on mine. I sweated profusely when I ate Rosa's enchiladas. Sinus congestion was no problem, but dealing with the flow of snot could be.

Another of the dishes Rosa prepared was tamales, but these were made mainly on special occasions like Christmas or someone's birthday. Once again, a part discarded by the rich was used—the head of a hog. Rosa would boil the head until she could remove all the meat and useable tissue from the skull, then mix it with spicy ingredients. She spread dried corn shucks with a layer of *masa,* made from ground corn, and then some of this spicy meat. She rolled these

---

* Even many modern Texans don't realize these *ristras,* or long strings of large dried red peppers, serve both as ornaments and as a useful way of preserving food. They are eaten by the native peoples of the Southwest. Before modern refrigeration and contact with other parts of the world, these chilies also represented the only source of vitamin C in their diet during winter months.

shucks containing this mixture and stacked them in a special pot with a small amount of water in the bottom. Rosa covered the pot and put it on the heat, and the tamales cooked in the steam.

It was not uncommon for Inés to invite people out on the weekends from the nearby town of Fort Stockton, some forty-five miles to the west. They always came loaded with beer, and on occasion, musical instruments. We spent more than one night serenading the stars and lost loves under the desert sky of the Bakersfield Valley, seated around a fire fueled by the hearts of long since dead mesquite bushes we routinely collected from the surrounding countryside. Inés was always drunk and I was always stoned.

* * * * *

Preparation for a cotton crop in the Bakersfield Valley began in the winter. José and Inés plowed, day after day, while West Texas winds raised clouds of gritty dust. Icy wind found its way through thick layers of clothing, and the constant vibration and noise from the tractor left their bodies numb and sore, but they persevered, one pass at a time, breathing through bandanas wrapped around their faces.

Next came the process of cutting rows, which were made with a tool bar armed with eight rows of listers. On either side of the tool bar, telescoping marker discs marked the ground to provide proper spacing for the next pass of the tractor. Any momentary loss of concentration would cause problems for the rest of the year.

After rows were laid out, Inés took a three-way blade and cut dirt ditches from which we irrigated the fields. We formed dams in the ditches using tarps, four-by-fours, pieces of sucker rod and plenty of hard labor with a shovel. Water was introduced to the ditches and then drained into individual rows with siphon tubes. Balancing the number of tubes with the output of the well was critical. Too many tubes would draw the water down until some lost their siphon and left dry rows. Not enough tubes and the water would rise until the ditch overflowed and probably broke. Each dam had to be reset every twenty-four hours.

We had as many as five of these setups going simultaneously, not counting those on the alfalfa and coastal fields. Irrigation engines failed and had to be replaced. We kept two spares on hand—the moment one failed, we replaced it. The disabled one went back to the shop to be overhauled or repaired.

After preplant irrigation, we planted seed. Some farmers put down a pre-emergence herbicide before this process, but we did not—perhaps because of the cost. The seed was sown into moist earth. Once again, great concentration was needed. If the rows were improperly spaced, cultivators and harvesting equipment would later be unable to do their jobs.

Once the plants emerged, we performed the first cultivation, and then we watered the field again. After each watering came another cultivation. Mexican laborers went through the rows with garden hoes to remove weeds missed by the cultivator and to make sure the plants were properly spaced. Constant vigilance was required to spot insect invaders which could make all of the above work null and void. When bugs were found, insecticides were sprayed either by tractor or by daring aerial pilots who zoomed just feet over the crop at full throttle, shooting skyward at the end of the field, turning sharply while almost straight up and then swooping back down in the opposite direction for another pass.

Watching the plants respond to all this care got into my blood. Each day the plants grew until the time came when acre after acre—450 of them to be exact—stood four feet high or better, loaded with bolls.

The banker came out and surveyed the field. "Best crop I've seen this year," he said.

Prices were holding steady. Our family owed the bank $800,000 at 14 percent interest, meaning we had to make over a hundred grand above all costs just to pay the interest.

Fall came and the bolls began to pop open. Now the once-green rows of plants sported spots of bright white cotton. In order to speed up the maturation process, we sprayed defoliant on the plants, once again from airplanes. We lined up a contract stripper to harvest the cotton. It rained. New leaves emerged. We sprayed again, this time with arsenic acid to kill the plants. The contract stripper assured us he would come on a particular day.

The date arrived, but he didn't. I drove by the fields, fighting off feelings of pride. The plants were dead but covered from top to bottom with beautiful white pillowy bolls of cotton. Never in my life had I seen such a crop. Cotton was bringing almost a dollar a pound, so we had a shot to pay down our loan and make a nice profit. Day after day, the contract stripper came up with excuses. We got pushed further and further back on his list.

Clouds began to brew in the evenings. I watched with fear and anticipation, but the storms held off. Finally the strippers arrived. The first day, they stripped twenty acres. It yielded almost three bales per acre, which meant we had close to $600,000 worth of cotton in our field. This money would knock down a significant sum from our loan, pay the damn interest and put us in good shape.

That night I watched as the clouds formed once again. Lightning raced across the sky and loud claps of thunder came closer and closer. A few drops turned into a steady onslaught of rain, wind and hail, buffeting our trailer house like a boat in a storm at sea. I knew without looking what I would find the next day, but I had to look anyway.

There was no difference between the rows that had been stripped and those that hadn't. Bare brown stalks remained; below was a mat of mud, leaves and cotton, laced with arsenic acid. An entire year's work was gone. If the cotton hadn't been poisoned, we could have at least turned cattle in to eat it, salvaging something.

José and Inés knew what else the storm meant—their jobs were on the line. No crop meant no money—money we needed to pay their wages.

* * * * *

Insurance companies play the odds, and they knew that there was a great risk involved in growing cotton in the Bakersfield Valley. Consequently the premiums they charged to insure a cotton crop there guaranteed a loss, even when a crop was made. So we didn't have any insurance.

* * * * *

About a month or so after my initial foray into Mexico, I asked José Chávez if he could set me up with a connection to buy some marijuana. José figured we were going broke farming anyway and decided this would be a good way to earn a few bucks with little personal risk.

One bright Sunday morning, we struck out for the border. José led the way in his pickup and I followed behind in my Suburban. I brought along several empty suitcases and a smattering of camping gear to look the part of a tourist. We drove from Fort Stockton to Alpine and then turned south and headed toward the Big Bend National Park. When we arrived at the park we turned east. We drove a few miles and then turned south once again, skirting up and over a rocky ridge that formed the eastern foothills of the Chisos Mountains. A spectacular vista greeted us as we broke out on the hill overlooking the river. The descent to the river was steep. The road wound back and forth, around and over colorful volcanic rock formations jutting from the earth. Vegetation was scarce—the few plants that managed to grow from the dry rocky ground were armed with thorns.

At the bottom of the road lay Castalón, an old cavalry outpost. We bypassed this and turned onto a road that led to the river crossing just across from the Mexican village of Santa Elena, Chihuahua. We drove through a densely wooded strip of ground subirrigated by the Rio Grande. A wall consisting mostly of salt cedar trees with occasional tall cottonwood trees and stands of *carrizos*, or river cane, engulfed us on either side. At the end of the road we arrived at a small parking area. We left our vehicles there.

I handed José the money to buy the marijuana and he disappeared on foot into the stands of salt cedar and river cane bordering the river. I remained on the American side and waited nervously. About an hour later, José reappeared and motioned for me to follow. He was dripping river water from the waist down. I followed him down a narrow trail and then into a thick stand of brush where a sack full of marijuana waited. Perspiration ran down his face. He looked around and listened intently for any unusual sound.

His hands shook as he handed me the sack. I looked over the contents. He turned, said goodbye and trotted off, almost in a run.

He left me standing there holding the bag—literally, without any instructions on how to proceed. There I was, proud owner of about twenty-five pounds of good quality Mexican marijuana, standing less than a hundred feet from the banks of the Rio Grande.

Now what do I do?

Since I was a total novice, I carried the sack back to a location near the truck and hid it in a thick patch of *carrizos*. I looked around to see if anyone was watching, removed the suitcases and carried them back to my hiding spot. I packed the marijuana into the suitcases, took another look around and then dashed back to the Suburban and threw them into the rear of the vehicle. No one saw me. I hopped behind the wheel, started the engine and drove toward Fort Stockton, almost sweating blood the entire way. Every oncoming car caused a wave of fear to rise in my breast and then an uneasy sigh of relief when my fear proved ungrounded. I watched my rearview mirror closely as well. But no one followed me.

I bought a bunch of ziplock bags at a grocery store, then rented a motel room. I dumped the weed into the bathtub. It filled it. I smoked as much as I could to stop from shaking and bagged up and weighed what remained for sale, culling perhaps a pound of the best-looking buds for my personal use. I repacked the suitcases and headed north.

Another hand at our farm had grown up in Plainview, Texas. He told me I could take the load to a childhood friend of his by the name of Arnold Kersh and that Arnold was capable of selling a lot of marijuana. He also called Arnold to forewarn him that I might show up.

Arnold and Jeff Aylesworth, a business partner of his, pulled into the parking lot in a huge old Fleetwood Cadillac. I watched them from a window as they approached my room. Arnold looked like many bull riders I knew—young, muscular, handsome and cocksure but not large.

Jeff was taller, a slender young man with long well-kept hair. Both had big intelligent blue eyes and exuded confidence—the kind women find difficult to ignore. Loud rock-and-roll music reverberated from Jeff's car. One look told me these men were young and wild as hell, not unlike modern-day James Dean types.

I got into my Suburban and followed them to Arnold's house in Plainview, trying to keep them in sight without being arrested for violating every traffic law known to man. This proved no easy task.

There was no weed to be had in Plainview at the time.

A day later, I left town with pockets full of money, somewhere in the vicinity of what I would earn in salary during a year of hard labor on the farm.

I was hooked.

DESPERADO

Over time, farmers—including us—switched to growing coastal Bermuda and alfalfa hay and grains, using cattle to graze the grain fields over winter. Chances of making a crop in the Trans-Pecos region increased with these crops, but both required a lot of water to produce good yields, leaving marginal returns at best. A natural gas plant was set up nearby and the once-free gas was now sold to farmers. That additional cost made it impossible to make a profit. Heavy pumping also diverted the flow of underground streams, and some of the wells began to produce salty water, causing salt to build up in the soil.

By the time I showed up on the scene, nearly all the original farmers of the area had given up farming. Most of us who did try to farm the soil were outsiders who thought we could somehow do a better job than our predecessors. We were mistaken.

Many miles separated us from markets and access to the parts and supplies necessary to operate a mechanized, irrigated farm. I spent a good portion of my time on desolate back roads—driving as fast as my truck would safely carry me—to pick up parts we needed, often discovering when I got back that something else had broken, and I would have to turn around again. The trip to Pecos, for instance, was ninety miles—one way. Alfalfa and coastal hay demand good prices in areas where horse farms and dairies exist. None of these operated near us, so high hauling costs ate potential profit.

Irrigating was never-ending, brutally hard, around-the-clock work, separated only by more work, cutting and baling hay. My hands developed callouses so thick they cracked and bled. We baled alfalfa hay at night to take advantage of dewfall to keep from losing the valuable leaves which would be pulverized into dust if the hay was baled too dry. Picking up the hay from the field once it was baled was also essential so the next round of irrigation could begin.

I worked between eighty and a hundred hours a week—and lost money. I can't begin to describe how frustrating it is to work so hard trying to grow the food we all eat just to go broke, even when successful in producing a crop.

Many have turned to a life of crime after enduring years of this frustration. One who comes to mind is a fellow by the name of Dick Graham.

My dad had hired Dick to help with the Bakersfield Valley farm during the years I lived and worked in Oregon. By the time I decided to return to Texas, their relationship had deteriorated and they had parted ways. Dick owned a neighboring farm and lived there with his wife, a son and two daughters. When I met him, his youngest daughter and his son were still attending high school in the closest town to the Bakersfield Valley—McCamey, Texas. Dick was the son of farmers from Iowa, who lived near the Missouri border. He had been a good farmer in his home state and had produced some of the highest recorded yields per acre of corn there, only to go bankrupt in the process.

Most farmers have learned to participate in every government giveaway program available in order to survive. To qualify for these programs, a farmer must relinquish many personal choices and comply with government wishes and demands. Don't comply and you get no help. Get no help—you go broke. Dick Graham found such a system contrary to his nature.

Dick subscribed to the belief that our government was controlled by a worldwide group of evil conspirators trying to gain control of us and rob us of our freedom. He saw the "system" as a rigged game full of catch-22s to keep us in line as slaves of these controlling power mongers. He believed that many institutions and their elite were part of that system, including banks, insurance companies, lawyers, and corporations. He saw government employees as pawns of these power brokers. He distrusted common, law-abiding citizens, considering them brainwashed, unknowing conspirators of the game. In short, he was pissed off at the world. He felt like he had done a good job as a farmer and had been robbed of his land and the just rewards of his efforts to produce food and feed for the rest of us.

Dick stood about 5'10" tall and weighed 185 pounds. He was trim, muscular and browned by many hours of exposure to the strong West Texas sun. When I first met Dick, he was a very unhappily married man. His wife was short, plump and involved in an ultraconservative Pentecostal church in McCamey. Dick wore a white T-shirt, with a pack of cigarettes rolled into one sleeve, and Levi's 501 jeans, which at that time were not so popular. Lace-up work boots protected his feet. A cap covered his head—it was usually cocked back, but if not, then riding low over his eyes. His hands were strong and calloused and his eyes intensely blue and piercing. Dick possessed a good sense of humor in a vulgar kind of way and he laced his speech with a colorful blend of cuss words. Most of the time he smiled. He smoked constantly and drank coffee at all hours of the day and night. He did not drink alcohol or take drugs other than his cigarettes and coffee.

Dick was afraid of no one. When he was a teenager, his father had hauled him from town to town to participate in no-holds-barred fights, gambling and making money off of his son's abilities.

After Dick had gone bankrupt in his home country, he mysteriously showed up in the Bakersfield Valley with enough cash to put a large down payment on a small, irrigated farm and to equip it for production with irrigation engines and machinery for both row crops and hay production. Considering that all of the farmers in the valley had already gone broke themselves, they were more than happy to sell him what he wanted without worrying about where the money came from. *Why look a gift horse in the mouth?*

Dick sold me a piece or two of farm equipment at below market cost. I didn't bother asking where they came from. And he didn't bother telling.

Dick, I learned, was highly educated in the art of *repossesin'*. Some might call this stealing, but that wasn't how Dick saw things. *The system* had taken from him and he was just taking back what was rightfully his to begin with. He didn't consider himself an indiscriminate thief. The main problem with Dick's view was that just about anybody could justifiably be included in his description as part of the hated *system*.

Dick liked his job as a repossessor and was very good at it. Since he was a talented farm mechanic, his specialty was repossessing farm equipment and tractors. He once told me that he could steal any brand of tractor made at the time with nothing more than a screwdriver. Dick was particularly fond of repossessing John Deere tractors. I suspect a look back at his past might reveal he once lost one or more to creditors.

When I would come into Bakersfield for parts, supplies or advice, I often found Dick slouched in a chair at the local farm co-op in the company of a couple of old-time retired farmers who owned and ran the store. Dick and these two men—Ralph and Walter—discussed all the world's problems and gladly dispensed advice to anyone who would listen. Their humorous observations often made me laugh. A favorite complaint was how the cost of everything related to farming kept going up while the prices they received for their agricultural products (grain, hay, cotton, cattle, etc.) remained the same or got worse year after year.

Sometimes they joked about solutions to this problem, some of which involved violent behavior. Only, Dick wasn't joking.

OSCAR, PRINCE OF THE NORTH

I first met Oscar Cabello through Vicente, one of the hands who worked on my dad's farm. Vicente was a rare breed: quiet, kind, considerate and reliable, with no discernable bad habits—he didn't drink, smoke, swear, womanize or do anything else that would have made him interesting or like the rest of us. He showed up for work on time every day and smiled a lot.

On the other hand, I had plenty of vices and wasn't overly concerned with hiding them from the world—except my marijuana use from my dad and a weakness for pretty women from my wife, which proved somewhat more difficult. All the hands knew I smoked pot and over time found out about my first aborted attempt to acquire marijuana in Ciudad Acuña. Shortly after José Chávez steered me toward a reliable source in Santa Elena, Vicente told me his oldest brother also knew people in the business who might be more honest than those José dealt with. I didn't really think much would come of it but told him to have his brother stop by sometime so we could talk.

* * * * *

A white Chevy van pulls up to the entrance of our farm and stops. I arrive from the field in a big farm truck, covered in mud and grime from irrigating. In the Chevy van sit two large Mexicans—or rather, one large Mexican and one huge Mexican. They're dressed in what I would call Western dress clothes with a Mexican flair—both wear boots made from exotic skins, expensive cowboy hats, fine-threaded Western slacks and Western-style shirts with snaps instead of buttons. The larger of the two wears jewelry: several rings, one of which depicts a golden bull with red rubies for eyes, a fancy watch and a thick gold chain bearing a crucifix around his neck. He is clean and smells of cologne. He introduces himself as Oscar. The other, a brother-in-law, he calls Chuy.

We shoot the breeze for a while, then Oscar asks me if I am interested in buying any marijuana.

I tell him that I am. I have doubts about his connections, simply because Vicente is such a meek and mild kind of guy, but I figure it can't hurt to check it out. Oscar does seem different—more confident and aggressive than his brother.

He briefly describes where we'll be going and where his hometown of Piedritas is in relation to the part of the river I know. I'm left with only a vague idea of its location. Oscar and I arrange to drive down to the river, where he'll sell me a small load, about twenty pounds.

Sunday morning arrives. Before sunrise, I meet him at Fort Stockton. He drives the white van, and I follow in my Suburban. We drive from Fort Stockton to Alpine and then head south through the open expanses of the Big Bend region. The country around Alpine is beautiful but it's hard to appreciate that beauty under these circumstances. I see broad grassy plains specked with cattle, presided over by brooding mountains welcoming the sun's first light. Jackrabbits dash across the highway. I try to avoid hitting them but their movements are unpredictable. Oscar doesn't bother trying to miss them. I hit just as many as he does.

We travel south. The scenery changes from that of grassy plains to creosote-covered desert with scattered bunches of ocotillo, lechugilla and cacti. Large volcanic rocks protrude from the ground as we near the park, looking like they've been cast upon the earth in random fashion or erupted from violent upheavals—the work of a mad artist.

We approach the base of the Chisos Mountains, an impressive range jutting from the desert floor to elevations approaching eight thousand feet, and turn east. About ten miles later we turn south again and head down River Road, skirting the eastern side of the Chisos. Here we once again encounter dry, forbidding land. Rocks and outcroppings jut from the earth. A layer of smaller rocks and pebbles covers the hardpan alongside, the apparent product of centuries of erosion caused by radical changes in temperature, wind and torrential rains that come only a couple of times of year. When they come, they're violent. The colors are impressive and the road semitortuous—it crawls up and over ridges that serve as footing for the nearby mountains and across dry, sandy arroyos that turn into roaring rivers when it rains in this otherwise arid land.

Oscar leads me all the way to La Pantera crossing, better known among gringos as Talley Crossing, or at least that's what the sign at the cutoff tells me. Midmorning finds us near our destination. The river is bordered by a thick strip of vegetation. We follow a path through a dense stand of salt cedar trees interspersed with large patches of river cane barely wide enough to allow our cars to pass. We pull up to the edge of the river and wait on an embankment overlooking the water.

Waiting doesn't come naturally for me. And knowing why I am waiting leaves me acutely uncomfortable. I watch. I listen. I smoke dope.

After a couple of hours, the sound of a pickup reaches our ears, approaching from the Mexican side. Oscar tells me to wait and disappears into the salt cedar trees. I never see the truck. Thirty minutes later—which seems like three hours to me—he reappears and waves, signaling me to follow. I follow the big man down a trail into a thicket where a cardboard box sits out in the open. I inspect the contents. The weed is brown and full of seed. I produce rolling papers. I break a bud, try to discard some of the seeds and roll a joint. Oscar declines to smoke with a smile and a wave of his hand.

"You don't smoke?" I ask.

"No," he replies, grinning. Almost shyly.

While I smoke the joint, a huge rattlesnake approaches. He follows the trail and we are blocking his path. He coils and begins to rattle, threatening us. This is the first and only time I've ever seen a rattlesnake confront a human without having first been disturbed. It proves to be a fatal mistake. Oscar calmly picks up a small tree limb and beats him to death.

The marijuana isn't great, but I think I can sell it. It's summertime and there's nearly always a shortage of Mexican marijuana during these months. We make our deal. I load the weed and leave. Oscar stays behind. I drive north—heart pounding, adrenaline rushing all the way home across an abandoned highway—and arrive without incident.

This is too damned easy, I think to myself.

★★★★★

This would prove to be the beginning of a long personal relationship—by dope trade standards—between Oscar Cabello and me. I would meet nearly all of his immediate family and come to know them like brothers.

Most white Americans have little justification for entering the drug business. While we may claim we do so to save the farm or our business or because no other options are available, the truth is it represents what we think will be easy money and a convenient way to stay high. We have other opportunities.

Some Mexicans have real reasons for getting into the dope business. Oscar was one of those. I won't say he used all the money he made for good things, because he didn't, but he did do *some* good for a hopelessly poor community with few or no realistic options for survival. The fact that his eventual arrest led to the collapse and demise of his community lends credence to this opinion. When I look back at the early life of Oscar Cabello, the line between what is right and wrong is hard to discern.

★★★★★

Oscar's father, Celerino, a little man with a huge smile, originally showed up in Piedritas, Coahuila—a small *ejido** in the desert mountains of northern Mexico just twenty-five miles or so south of the border and very near the line that separates the Mexican states of Coahuila and Chihuahua—after having done something illegal in another part of his country. A good many of the residents in that region arrived under similar circumstances. The area in which Piedritas lies is known as the *despoblado,* or badlands. Piedritas is probably the most remote of all the villages there—several hours of rough dirt roads separate it from the nearest pavement. Celerino married a woman from an influential family in Piedritas, the Villarreal family. Oscar was the first child of the union and the oldest of three boys and several sisters who would follow. Unlike Celerino, Oscar grew to be a big man, favoring the European blood of his mother. When I met him, he stood several inches over six feet tall and weighed close to three hundred pounds.

Celerino, like others, tried to raise his family around Piedritas and discov-

---

* An *ejido* is a parcel of land actually owned by the Mexican government, but the right to use it is shared by the *ejiditarios*. Established after the Mexican Revolution, the *ejido* system was meant to replace the hacienda system of land ownership. The haciendas were huge tracts of land owned by *hacendados* who ruled over their land and the people who worked on it like feudal land barons. In theory, an *ejido* is a democratically operated communal farm and/or ranch operation. However, the reality seldom resembles the theory. The word *ejido* can be used to refer to the land governed by the people or to the small community, like Piedritas, where the people live.

ered the impossibility of doing so by any legal means available to him. He ran a few cows and goats, grew corn, beans, tomatoes, chilies, melons, a little sorghum, and alfalfa—and almost starved. Needing money, he decided to try his hand at producing and smuggling an illegal product to the gringos—*candelilla* wax. This valuable wax is derived from candelilla, a spineless, fibrous succulent that grows in clumps of pencil-thin, greenish-gray protrusions. During the early and middle years of the last century, the wax made from the plant was valuable as an ingredient in chewing gum, cosmetics and other products available at the drugstore, and it was even used as a lubricant in the manufacture of bullets. *Candelilla* wax could not be sold without a permit from the Mexican government and no permits were available to those around Piedritas.

The plants dot the hills surrounding Piedritas. Locals take burros to these hills and pull the clumps from the ground by hand, stack them on the backs of their trusty beasts of burden and haul them back to a *paila.* Pulling these clumps from the soil is difficult, leaving both back and hands sore from exertion. Finding good sources of *candelilla* requires a lot of walking over rough terrain. The *paila* is a large, rectangular metal vat placed in the ground and filled with water. It has to be placed near a source of water, a scarce commodity in that part of the world. A space is left open under the *paila* where a fire is started. Clumps of the *candelilla* are added to the hot water along with a quantity of acid. This causes the waxy coating of the plants to separate and float to the top of the vat where it is skimmed off and accumulated. The wax then dries into hard, light brown chunks with special inherent properties that give it value.

In spite of all these difficulties, Celerino harvested the plants and produced the wax, made money and soon found himself in a position of leadership in the community. Then one day, *forestales,* the Mexican equivalent to our game warden or park ranger, showed up and intercepted Celerino and some of his men with a large pack train of around fifty burros laden with *candelilla* wax. The load represented six months of backbreaking work. Shots were exchanged and one of the *forestales* lay dead. Celerino and his men took off with the load and evaded capture. Celerino was blamed for the killing.

For five years, he lived in Piedritas and successfully avoided all attempts by the Mexican government to apprehend him. At some point, the crime was either forgotten or forgiven and Celerino became the president of the *ejido* (equivalent to a mayor in our country). He built a church and took it upon himself to care for those who could not care for themselves, using profits from the illegal *candelilla* wax trade.

The poor residents of communities like Piedritas have no welfare program and none of the benefits Americans enjoy and often take for granted. When someone without family gets too old to work, they do without. Retarded people without caring family members do without. Same for cripples or any other handicapped person. Those people soon learned that Celerino would provide for them—perhaps not on a grand scale and not without some prejudice, but he did provide when no one else would.

The day came when *candelilla* lost most of its value—an entire day's worth of backbreaking labor scarcely yielded five dollars. Likewise the once-profitable business of selling sotol and tequila died as Americans found easier ways to obtain alcohol legally. People around Piedritas dug fluorite out of the mountains by hand, earning in the vicinity of five bucks a day for a brutal day's effort. More than one man remains buried in a deep mineshaft which collapsed and crushed him. Others poisoned themselves digging cinnabar and extracting mercury destined for the United States, a practice which also yielded little money. Children suffered from curable diseases like a virulent form of conjunctivitis which leads to blindness, where a dollar's worth of antibiotic ointment would have saved their eyes. They suffered and died from dysentery caused by drinking out of contaminated water sources. There were no sewer systems. And nobody on either side of the river gave a damn.

This is the world Oscar was raised in. Maybe because his dad did a better job of feeding him as a child, Oscar was more intelligent than his peers or even than his own father. Not only was he more intelligent, but he was bigger, stronger and more compassionate. Because of this, he inherited the burden of taking care of the needs of his community at an early age. He was elected their representative in the state legislative body that ruled the area, but that did little to bring aid to the community. Piedritas had nothing those in power wanted, and consequently they had nothing to offer Piedritas.

A few aborted attempts to help did take place. A water tower was built but the water system that should have accompanied it was not. So the tower stood rusting off in the distance, like some huge monument to the goodwill of the Mexican government, while women and kids pulled water to the surface from a well in the center of town, using ropes and buckets. A large dam and gravity flow system was built to collect rainwater for irrigation purposes and a lake-full of water accumulated. For a few years crops flourished, but the government screwed the farmers out of what they should have received. The people had no machinery to work the soil or harvest their crops. The Mexican government provided these things. Then, when the crops were harvested, they paid the people with a few sacks of flour and kept the money. The *ejiditarios* became disillusioned and quit.

Oscar made contact with the rich world of the United States at some point in his young life and discovered there was something Mexico had that the people of this country wanted and would gladly pay for: marijuana. Not only would they pay for it, they would pay a lot for it. Oscar didn't have any interest in getting high, but he did need money—desperately—more than he could earn working legally in the United States. He found the sources, and then American buyers found him.

Celerino had reservations about entering the marijuana business. He had heard all the propaganda but remained undecided about whether it would be a bad thing. Oscar entered the business nonetheless and gained his father's favor by investing profits into legitimate businesses for the community: cattle, horses, pick-up trucks and a limited amount of farm equipment. In time, nearly the entire village worked for the Cabello family—farming, ranching and smuggling marijuana. Those who didn't starved, trying to make a living collecting *candelilla* or digging fluorite. Celerino fed the old, retarded and incapacitated of the town and paid a traveling doctor to come by on occasion.

Oscar began to move larger and larger loads through the region, and he made more and more money. Then came setbacks. One man, who had always been reliable, simply drove off with a ton of marijuana—bought on credit—and disappeared, never to be seen again. Oscar had to make good on the debt to those higher in the food chain who were supplying him with his product. If he didn't, the results could be tragic.

Then came the ill-fated shootout at the San Vicente crossing described in Terrence Poppa's biography of Pablo Acosta, *Drug Lord.* An American narc set up a deal to buy a load of marijuana. The van to be used to haul the load concealed American drug agents. The agents jumped out after the Mexican vehicle containing the load arrived on the U.S. side. Shots were exchanged during the attempt to arrest the men. The Mexicans fled, leaving the marijuana behind. While this may be attributed to Acosta, it happened on Oscar's turf, and he suffered much of the financial loss since most of the marijuana was his. Oscar had nothing to do with any shooting that day. Perhaps Acosta's people did. In any event, Acosta was inclined to take credit for what happened, and smarter men were more than glad to let him.*

It was about this time that Oscar began to look for new buyers. He needed someone he could trust in the United States. And I needed someone I could trust in Mexico. Vicente connected us. I had no money to speak of, but Oscar was will-

---

* According to Poppa's account in *Druglord*, David Regela, a U.S. Customs agent, claims to have killed a number of Mexicans during the shootout. According to my sources, only one died that day—the government informant who set up the deal.

ing to front me the dope. A man with dope will soon find those willing to pay for it in this country. Matter of fact, they'll find him.

I did not want to sell to West Texas locals. To me it seemed like shitting in my own bed. Besides, no one out there had much money. Instead I approached Arnold Kersh and his crew in Plainview, who could supply the Lubbock area. Later I also contacted my cousin Phil who lived in the Dallas-Fort Worth metroplex. So Oscar bought from his connections in Mexico, he sold the marijuana to me, and I carted it across the river and sold it to Arnold, Phil and others. Their job was to collect everybody's paychecks. Oscar—and the rest of us too—started to make lots of money.

All of this wealth eventually took its toll on Oscar. While he did continue to turn over healthy amounts to his dad for distribution into the poor community of Piedritas, he also developed a liking for cocaine and whores. It seems to go with the business. Oscar moved his family to Fort Stockton, Texas, and bought a small conservative home. His children learned English and enjoyed the privileges of living in the United States. Oscar began to spend more and more time gone from home in the company of people like Amado Carrillo and high-ups in the Mexican government—drinking brandy, snorting lines and screwing. No matter how much money a man makes, there's never enough for that lifestyle. Maybe there're days when, to him, the cash seems unlimited, but mark my words—the day comes when he'll look up and it's gone.

Over time I, too, would fall into the same destructive behavior traps. Maybe it was because of the tremendous amount of stress marijuana-smuggling brings and maybe it's because of the money that opens doors. My marriage became a nightmare—Cheryl and I fought constantly so I'd just drop off money and leave. Whores seemed better companions than "good" women because they didn't ask questions. You paid for what you wanted without having to get involved in their lives. At least that was the way it was supposed to work. Only problem I had, I fell in love with nearly every whore I screwed and then found myself supporting her. I ended up with one in Lubbock and another in Fort Worth before my game came crashing down.

If you have five hundred dollars, the cost of a whore is five hundred dollars. If you have five million dollars, your whore will cost five million dollars. Take it from me—the cost of a whore is all you have.

And down in Piedritas, some little kid goes to bed hungry and sick.

ARNOLD, A SHOOTING STAR

Arnold Kersh was blond-haired, blue-eyed and handsome, with looks not unlike a young Brad Pitt. He attracted girls by the droves. The rest of his friends hung around trying to get high for free and seeing if they could snag one of the left-over women. Arnold believed in sharing.

Arnold's house actually belonged to his parents, but they were always gone so he had the run of the place. Pills of every description sat on a coffee table—free to all, it appeared. Alcohol flowed liberally through the door and into the bodies of people who congregated all hours of the day and night. During the brief times I was there, I watched from the sidelines as money walked in and smile-wearing, marijuana-toting people walked out.

While Arnold nicknamed me "Wild Man," the name suited him much better. Arnold's favorite state of mind was what he termed a "high-speed wobble." He arrived there by getting high on very large amounts of crystal methamphetamine which he combined with downers and topped off with generous quantities of alcohol and an endless succession of joints.

Most of Arnold's friends were cowboy types and exhibited the common traits: boots, a big buck knife in a scabbard hanging from a hand-tooled leather belt, Wranglers with the imprint of a can of Copenhagen in one rear pocket, and cowboy hats. They drove pickups with a splattering of tobacco juice on the driver's side, huge mud-grip tires and a gun rack in the rear window. But they delivered drugs. *Cowboy mafia.*

They partied all night, every night, and slept during the morning hours. They liked "outlaw country" music, people like Waylon Jennings, Johnny Cash, David Allen Coe or Jerry Jeff Walker. It was not uncommon to hear loud rock and roll blasting from blown-out speakers in their houses or pickup trucks—the hard stuff like AC-DC, Black Sabbath, or Led Zeppelin.

For fun they played pool, chased women and occasionally crawled onto the back of a bull to see who could hang on the longest. Any highway sign they drove by was liable to get hit by a spent beer bottle or a salvo of bullets—or it might get run over. In this crowd, driving the speed limit was forbidden, and there was a race to anywhere they happened to be going. Passing through the ditch or even a bump here and there was considered fair play.

They also liked to fight, but in that era fighting rarely included guns or knives. The resulting injuries included black eyes, broken noses and various bruised or broken body parts, but never resulted in anyone's death. Arnold was particularly well known for his abilities as a fighter. Though not a large man—he stood about five-ten and weighed maybe a hundred and sixty or seventy—Arnold was strong, quick and athletic, with lightning-fast reflexes. And he was strong-willed. I never saw him whipped.

On one occasion a large cowboy-type, weighing two hundred pounds or better, called Arnold out in a bar. Arnold stepped up to the larger man, removing his own jacket as he did. The big man began to remove his as well but before he got his arm out of the sleeve, Arnold yanked the man's coat down, trapped both his arms, and then proceeded to pound his unprotected face to a bloody pulp. The only injury Arnold received was an infection from the big man's teeth which got imbedded into the knuckles of Arnold's right hand.

A wide variety of people bought drugs from Arnold. Plainview's economy was mostly agricultural. Hale County, in which it resides, produced more grain crops than any other in the state and rested above a large aquifer which unfortunately was in the process of being sucked dry. A surprising number of these farmers plowed all night while high on crank or cocaine, perhaps with a beer between their legs and a joint thrown in for good measure.

A large slaughter plant nearby employed lots of blacks and Latinos. So did cotton gins, grain mills, silos and other support businesses for the agricultural community. Each of these businesses provided money that ended up in Arnold's pocket and those of others like him. He also accumulated material offered in trade for drugs—things like stereos, televisions, guns and vehicles. His freezer was usually full of prime steaks and meat which walked out of the Missouri Beef packing plant and provided the material for frequent cookouts.

And then there was Lubbock, a short drive to the south and home to Texas Tech University and a bunch of college kids with money Mom and Dad sent to pay bills—if their parents only knew. Lubbock was a "dry" county, heavily influenced by ultraconservative Christian doctrine. To get beer, these students had to drive to a nearby town, which wasn't a town at all but just a couple of beer and liquor stores that moved the stuff by the semi-load. I always found it amusing to see the guy who owned and profited from these liquor stores out pandering to the "Christian vote" when repealing these laws came before the public. During this era, marijuana ran a close second to alcohol as the drug of choice among college students, and Arnold did what he could to get his share of that market.

However, Arnold's primary reason for being in the business was not amassing money. He was on an endless pursuit of a good time, aided and paid for by drugs. He shared all he had with his friends.

I avoided participating in the activities of Arnold's crowd, fearful that word would leak about what I really did for a living. He usually introduced me as an old friend who just happened to be passing through. I always came loaded with marijuana, pre-packaged into pounds, and picked up bundles of cash for the previous load. I sold him the marijuana on a credit basis and had no trouble collecting the money he owed me. I arrived unannounced and rarely spent over a couple of hours in town. He liked the idea of keeping others in the dark too, knowing that if they knew I was his supplier, they would try to go around him to get the product cheaper. He made a handsome profit on all the marijuana he sold. I'm sure that eventually some people knew what I really did, but I refused to discuss business with anyone but Arnold, and the rest were afraid to approach me on the subject. Arnold and I became good friends.

Arnold brought a whole new meaning to the word tough. Like Dick Graham, Arnold feared little in this world. I once saw him choke a pit bull unconscious with his bare hands after the dog made a threatening move. Another fellow told me Arnold once found himself hemmed against a fence by a mean old cow while working in the pens of a livestock auction. It is said he managed to insert his hand into her eye socket, grab her by the eyeball and back her across the pen, cussing her all the way. There were things in our world Arnold should have been afraid of—but then, Arnold wouldn't have been Arnold.

★ ★ ★ ★ ★

One winter day, February 25—my twenty-fifth birthday, to be exact—I drove into Plainview with a load of marijuana. Police cars surrounded Arnold's house. My vehicle was full of marijuana so I didn't stop, opting to call from a pay phone instead to see what was happening. To my surprise, Arnold's mother answered. This was the first time I ever talked to her.

She told me he was dead.

I'll never forget my twenty-fifth birthday. I spent it in a state of shock and agony, grieving over the loss of my good friend. Even now, thinking about it plunges me into a state of mind hard to describe.

It was a long time before I found out how Arnold died. I still don't know for sure, but this is what I was told:

A man commonly referred to as Buzzard sold Arnold some crystal meth he got from a group of bikers out of Lubbock. Buzzard was a heavy user of the drug and removed a lot out of what he got for Arnold, replacing it with cut before delivering it. By the time Arnold got it, it wasn't any good. Arnold refused to pay.

The bikers told Buzzard to collect. Of course he couldn't tell them he had stolen half or more of the product before he delivered it, so he just told them Arnold refused to pay.

Buzzard showed up one day at Arnold's house, supposedly to reconcile their differences. They went into a back room together. A few minutes later, Buzzard came running out of the room shouting, "Hey man, something is wrong with this guy." Other people in the house went back into the room and found Arnold lying on the floor. He had pissed on himself. His heart had stopped beating. They couldn't save him.

An autopsy revealed a needle hole in his arm. Anyone who knew Arnold also knew he had a fear of needles and never shot drugs. Other people in the room speculated that Buzzard gave him something to snort, which knocked him out, and then shot a drug into one of his veins. I don't suppose I'll ever know for sure. Less than a month later, someone shot Buzzard between the eyes with a .44 magnum. The police didn't waste much time investigating the crime once they found out who the victim was. None of this brought Arnold back to life.

The once joyful face of my friend was replaced with the haunting memory of his body, lying in a wooden casket, dressed in Wrangler jeans and a white Western shirt. I wanted to ask him what had happened, but the dead don't speak. He never saw *his* twenty-fifth birthday.

ANNA, THE WHITE ROSE

Arnold's friend Jeff Aylesworth moved a substantial portion of the marijuana I fronted to Arnold. So after Arnold died, I looked Jeff up. I needed someone I could trust to take over Arnold's share of the business.

Jeff was in his early twenties but looked seventeen, with long well-kept hair and a tall thin-framed body. He had an aversion to real work but did well hustling others out of their money playing pool. Jeff had spent a lot of time growing up in a billiards hall his granddad owned. He learned to shoot pool at an early age, standing on top of a milk crate.

Bars served as his office, but most of the locals knew better than to gamble against the guy. In order to make any big money he had to travel. Jeff drank a lot and smoked Marlboro Lights. He also smoked a lot of marijuana, swallowed downers and pill speed, snorted crank or cocaine and occasionally ate a hit of acid. He didn't shoot drugs—into himself, anyway—and I don't believe he ever tried heroin though I wouldn't swear to it.

Women found Jeff attractive and he had a slew of them scattered around, ready to lay him down whenever he got the urge, and he got the urge quite routinely. Jeff had a lot of connections when it came time to sell his dope, and most deals were closed over a pool table. Arnold had been able to keep Jeff in line. I found the task near impossible.

Jeff had an engaging personality. He could talk most of the women he met in bars out of some "booger," and most of the men in the dope business out of their dope, me included. I normally sold him marijuana on credit. He usually sold it quickly but if I wasn't around to collect immediately, he would reinvest the money into someone else's pot or whatever drug happened to be going around at the time. When Arnold was alive, I rarely waited for my money. I was in and out of town and hardly anyone there saw my face. With Jeff, I arrived—and waited—and hardly ever collected all I was owed.

If you can't take waiting, the dope business is not for you. When I waited on Jeff, I spent days locked in a cheap motel, smoking one joint after another—after another. I didn't like hanging out with Jeff much—that would expose me to all the locals and soon the word would get out. So instead of staying in Plainview, I usually rented a room in Lubbock.

One day, out of boredom I went to a tit bar owned and operated by the Bandidos, an infamous biker gang. While I was there, I spotted a big-boned, blond dancer by the name of Anna.

The bar was called the Cheyenne Social Club. Anna's sexy figure caught my eye. I filled her G-string with money and she found me at a table shortly after her time on the stage. In a matter of minutes, any resistance I might have had melted.

Anna stood about 5'8'' tall with the loose curls of a permanent. She was a natural blond, but not near as blond as the peroxide made her. Her body was muscular but still feminine. Her eyes were ice blue, a gift her Russian mom gave her, her skin tan, thanks to the Italian blood her dad contributed.

She told me she had been raised in Ohio but left home at an early age. Bikers found her at a truck stop and had taken her in. After a time, the original biker who discovered her sold her to DJ, her present owner. Bandidos owned the club and controlled the local meth market as well. At the time, they were probably the second largest bike club in the country—second only to the Hells Angels. DJ was the National Sergeant of Arms. I figured I might try to approach him through Anna to sell marijuana to the group. Or at least that's one of the ways I justified seeing her.

Anna told me she was a working whore, so one evening I bought her for a night. The price was five hundred bucks and most of her clients felt like they got their money's worth. Rather than bed her, I took her out to eat. I did this several times. Every time, I paid for her services and then sent her home without having sex.

DJ found this perplexing. One evening he picked Anna up after our date, just to introduce himself. I guess he was feeling me out to see if I was a cop. DJ was at least six feet tall with a muscular build. He had dark eyes and dark hair which he did in a ponytail that reached the middle of his back. He wore black leather and lots of jewelry, mostly skull-and-cross-bones kind of stuff. If he intended to scare me, it didn't work.

Anna and I grew close. I was torn. On one hand, I hadn't yet run around on Cheryl and didn't want to. On the other hand, I did want very much to have sex with Anna. The day came when the latter hand won that battle. We spent an entire night together. From that day forward, I was infatuated with Anna. I suspect the feeling was mutual.

I also began to sell marijuana to the Bandidos, more for their personal consumption than anything else. At the time I had no idea they might be connected to Arnold's death.

* * * * *

Out of frustration with Jeff, I eventually called my cousin Phil to see if he could move any marijuana. He smoked dope and lived in Fort Worth. The Dallas-Fort Worth metroplex had a lot of people and a lot of that number liked marijuana. People there earned good money. My cousin Phil had worked at a variety of construction-related jobs over the years and had established sources for marijuana everywhere he went. Becoming a dealer instead of a consumer was easy for him. He simply went to the people he bought from with a better deal than they could get from their own connections and then worked backward through the various levels of suppliers, eventually finding the major wholesalers in the city. We were close enough to the source to beat or match anybody's price and still make a handsome profit.

Marijuana is produced seasonally, but the season for smoking the stuff never ends. When a supplier has nothing to offer dealers, they'll take the money they owe him and go somewhere else. Inevitably a large part of our uncollected money never got paid at the end of the season, but I always had to pay my debts or risk ruining my connections. Maybe the reason people like Oscar and me never rose to the level of an Acosta or a Carrillo is that people knew they could rip us off and we wouldn't do anything. Acosta and Carrillo would kill you—and maybe your family as well. They always got paid, if not with money, with blood.

BALMORHEA, THE DEVIL'S SWING

The day arrived when I received word that my dad's farm was going to be shut down. For a time the bank wanted every dime we generated, yet wouldn't allow us money to pay the hands, buy parts or even the wire we needed to bale hay, all of which would generate more money. It became obvious that they were only interested in closing us down and selling off what we had.

I didn't take this news well. For a time I paid the hands with cash generated from the sale of marijuana. At one point, I considered hauling off all the equipment and selling it in Mexico but better reason prevailed. I had the awful responsibility of firing the hands. Then I decided if they were going to be fired, so was I. I left.

I moved my growing family, now consisting of three boys, a girl and a pregnant wife to the small town of Balmorhea, about a hundred miles west of Bakersfield. In most parts of the country, such a distance would mean a new state. In the Trans-Pecos region, there was only one town between Bakersfield and Balmorhea—that town was Fort Stockton. Residents in that area consider people from all three towns neighbors of sorts.

I had hopes of making a lot of money smuggling and wanted to buy a farm at Balmorhea. The area appealed to me for various reasons. First Balmorhea was much prettier, nestled against the northern slopes of the Davis Mountains. Second, and more important, it had a natural spring that flowed millions of gallons of water a day, most of which ended up in a ditch used to supply water to the town and the surrounding fields. This water, coupled with the rich soil in the area, created an oasis in the desert, with trees and shade not to be found anywhere else in the region. The largest impediment I had found in making money growing crops around Bakersfield was the cost of pumping water. While land prices were higher around Balmorhea, I figured it would be worth the difference, especially since I hoped to buy this land with dope money and wouldn't have to worry about repaying a bank.

My plan was good except for one problem: my move coincided with one of those seasonal droughts in the supply of Mexican marijuana, and I had virtually no money nor credit to buy land. Rent was cheap though, so I rented a house and waited, knowing the drought would end and I would have another shot at making money smuggling marijuana.

Balmorhea was poor, with more than its share of needy families. The entire population consisted of about six hundred souls—half of them were unemployed by the traditional definition of the word. As a rule, when one of the poorer citizens in Balmorhea prospered, all the rest did as well. When one did without, it was likely that all did without. Unfortunately, the latter condition was the more prevalent. Over time, Balmorhea's residents had learned to share what little they had to survive. This went for marijuana too.

When one *marihuano* had marijuana, chances were good that all the *marihuanos* living there got to smoke. At least one person always figured out how to get his hands on some. People shared their stash—maybe not entirely out of generosity—because they knew that when the day came that they had none, someone else would provide for their need as well.

I don't know how dopeheads spot each other, but I know we do. Within twenty-four hours of my arrival in town, I met Mathew Walker, another *bolillo*[*]—a white guy like me. He lived next door to the house I had rented. He had a wife, Amy, young and redheaded, and two infant children. We were getting high together before the sun went down.

There was no apparent means of support in the Walker household. But Matthew always had a little cash in his pocket. He was a big burly cowboy-

[*] Literally, a bolillo is a hard crusted white wheat roll ubiquitous in Mexico. Mexicans use the word as one more pejorative slang word for gringo—hard on the outside, soft on the inside.

looking guy. He always wore a hat and a thick reddish-tinged handlebar mustache. He usually had some degree of beard stubble, but he never cut off the mustache and fooled with it constantly. Mathew liked beer and smoke, and he liked his fighting roosters. He had good fighting roosters. We'd sit around smoking marijuana, exercising his birds, occasionally watching his cocks spar with gloves covering their spurs—and we'd plot against the white folks. It was through Mathew that I met the rest of Balmorhea's *marihuanos* and he gave me a rundown on each as they came along.

Steven Garcia, or *Cubano,* was the guy most of the young women found attractive. The day would come when my wife, Cheryl, would be counted among them. He was lean, athletic, had a great smile and was born with the gift of gab. He came along near the end of a long string of kids numbering in the teens and had to fight for his share of attention in this world. But something was lacking in Cubano—he couldn't seem to love deeply or make any relationship last. He seemed doomed to be the guy the girls all want to sleep with yet none want to keep. He was a gifted rocklayer. Sometimes he'd get a job rocking someone's house or building a barbecue pit. When he did, he employed the rest of us to mix mortar and to collect, haul, break, and pass him rocks.

Leroy Hernández was a good-hearted guy, but unmotivated. He was the last son of an extremely old and wise man and was perhaps a shade or two more intelligent than the rest of his crowd. He was heavyset and took life at a slow, comfortable pace. No problem ever seemed to upset Leroy much. I spent a lot of time in Leroy's company listening to his dad describe the early days of a working cowboy in the area with a wistful faraway look in his eye.

Robert Ortega, or *Crocket,* was an alcoholic Vietnam vet who received a monthly check for injuries he had received in the war. He always wore a big hat with an eagle feather in the hat band. If anyone asked, he would say it was a turkey feather to keep from being fined or imprisoned. Aside from being shell-shocked, totally insane and drunk all the time, he was a blast to be around. He was capable of just about anything and prone to keep me guessing what might come next.

There were others as well. Each deserves mention, if for no other reason than the way they took care of each other in times of need. Goo, Tato, Norman, Martíin, Lupe the *Indio,* and Sammy the Lowrider come to mind. Yes, even Balmorhea had one lowrider.

Not only did these young men keep each other high and drunk, they also shared their food, shelter, laughter and tears. While generous with each other, they'd steal a rich man blind and were not thought much of by prominent members of society or by the law enforcement agencies designed to protect these rich men and their possessions.

A look back into the history of the area provides insight into how they developed these traits. Most of these men had at least some Native American blood in their veins and those who didn't sympathized with members of that heritage. It was said among these natives that outsiders, both in Washington, D.C., and in the *Distrito Federal*, had stolen their land and their inheritance. Consequently, over the years they developed a double standard of morality where stealing was concerned, finding it much easier to steal from those who had first stolen from them.

For many of the people in Balmorhea, selling booze and other illegal products—produced by close relatives just across the thin stream of water separating our countries—was considered honorable work. Smuggling was a natural part of their lifestyle. The products changed over the years, but the work stayed the same.

While the ruling class viewed smugglers as evil criminals, poor Mexicans and Mexican Americans on both sides of the river held smugglers in high regard. They were heroes—modern day Robin Hoods, acquiring and redistributing wealth to areas left out by those who controlled the money and the power. The songs they sang and the *norteño* music they listened to glorified the exploits of smugglers. The gun battles the smugglers participated in became legend. If you hear the accordion-accompanied wail of a Mexican ballad, chances are good that the song will be about Pablo Acosta or Fermín Arévalos or Amado Carrillo and how they fought the establishment, killed their enemies, and died with honor. And their loss is lamented.

In spite of the fact that I was a *pinche gringo,* these men saw me as an outcast of my own race, a man who ate the same food, spoke their version of Spanish and understood their plight in the world. They welcomed me into their community.

A TV was worthless around Balmorhea without cable or a satellite dish, both still very expensive during this time. It was over a hundred miles to the nearest television station, so a good bit of our free time was spent outdoors, cooking, eating, getting high and getting drunk, all the while sitting around and telling lies. I myself spent a disproportionate amount of time getting high—so much, in fact, that there was little time to drink and what drinking I did was done primarily to wet down the cotton-mouthed condition smoking marijuana creates. The marijuana also gave me the munchies. In Balmorhea, we ate both kinds of food—Mexican and barbecue. But the original Mexican style of barbecue had been forsaken for the method introduced to Texas by members of another oppressed minority—the black man. In old Mexico, *barbacoa* is made by cooking the entire animal, which is wrapped and placed into a hole in the ground above a bed of hot coals and then covered and left buried under another fire for perhaps twenty-four hours. On rare occasions, like someone's birthday, this was still done around Balmorhea.*

The black men of Texas took pieces of meat the white men did not prize—namely the ribs off of a hog or steer, and the brisket, which comes from the chest area of a beef—and found a way to make them better-tasting than prime steak. Now this method is the accepted version in virtually all of Texas. Cooking good barbecue this way requires a special pit, or wood-burning oven, which allows the meat to be cooked far away from the source of the fire over an extended period of time.

Some of the men around Balmorhea had pits. But, more often than not, they cooked over an open grill made from native rock and enclosed on three sides. The grill would be placed several feet above a wood fire, using mesquite hearts. This and pecan were the only suitable woods readily available in the area. Ribs were easy to cook by this method. But a thick brisket was more difficult, requiring an entire day's roasting to get tender. After several hours on the open grill to absorb the flavor of the wood smoke, the brisket would be wrapped in multiple layers of tin foil to cook until tender. I sometimes cheated and finished mine off in an oven.†

Properly cooked, a brisket should have no sauce added during the cooking. Some people use a dry rub to add flavor. Others use nothing but a little salt and pepper. Still others add nothing at all. Time and clean-burning smoke create the product, and there is no substitute for either. The worst enemy of a brisket is black smoke created by incomplete combustion, usually emitted while trying to choke down a fire. This black smoke sticks to the meat and gives it a bitter bite, sure to cause heartburn.

Another item which occasionally found its way onto the grill around

---

* What now passes for *barbacoa* in Texas is the meat taken from the cooked head of a beef, usually baked in an oven.

† A brisket cooked in an enclosed pit without the use of foil produces a superior product. Few folks around Balmorhea, however, had an enclosed pit, nor would they have known how to use one if they did.

Balmorhea was the body of a baby goat, commonly known as *cabrito.* I don't much care for the practice of slaughtering baby animals, but I must confess that I have tasted *cabrito* and liked it.

On occasion, someone killed a hog. When this happened, the whole town shared in the event. This custom began before modern refrigeration was available and endures to this day. Since a full-grown hog is large and can't be consumed in one sitting, residents shared with their neighbors and expected the same in return. *Carnitas, chicharrones, carne asado* and tamales were the principle dishes derived from a hog killing, along with the lard produced as a by-product.

Fresh homemade tortillas and hot peppers of one type or another always accompanied these meals. When available, green chilies were roasted and—if any were to be had—avocados served as a delicious addition, sometimes eaten in pieces with a little salt and at other times mixed into guacamole.

Plenty of marijuana and beer helped it all go down a little better. Somehow, the surrounding poverty seemed easier to bear in that state of mind.

* * * * *

Once I reestablished my contacts in Mexico, I carefully avoided selling any quantities of marijuana on a local level. But I know a lot of them figured me out because I was never without for long. I would go to the ends of the earth to ensure I was never without. And naturally, since I always had marijuana, I was always invited to gatherings. If not—then the gatherings came to me.

THE ROSE HAS THORNS

On a routine trip to Plainview to deliver marijuana to Jeff, I stopped by Lubbock and the Cheyenne Social Club to discover Anna was gone. Another of the girls told me she had been sent to work at another bar in Lubbock which was also owned by the Bandidos. I went to the second bar and found her. Anna was distraught. She told me the Bandidos had murdered a dancer by overdosing her with drugs. I had seen this girl dance before—she, too, was beautiful. Local police called the murder a suicide, but Anna was adamant that it was a murder. She wanted out, but was afraid to leave.

I offered to take her out of there whenever she got ready. She asked me to come the next day. The following day I returned to find her bruised and beaten. I told her to change her clothes, gather her things and meet me at the door in exactly fifteen minutes. She was afraid, but I convinced her to try. I told Anna I was going to walk out the front door as if I was leaving, but when she exited the front door, I would be waiting.

I walked out of the door, got in my car and waited, checking my watch. Then Bandidos began to arrive on motorcycles, two by two, about every thirty seconds. All of them were dressed in black leather. Most had long hair and were covered with tattoos, but there were a few bald ones in the bunch. Their bikes were polished. Each pair pulled up to the intersection in perfect unison—like that of a marching band or a column of soldiers. I guessed DJ, Anna's owner, suspected what was about to happen and figured he'd present a little show of force to make sure I didn't take her.

While I waited outside, the procession of bikers continued, two by two, until there were maybe fifty or more in the club. Then it was time. I drove to the door. Anna stepped out of it and into my waiting car. We sped away. The only thing going faster than my car through Lubbock was the adrenaline-laced blood roaring through my veins.

I wasted no time getting out of Lubbock. That night found us in a motel room in Odessa. Anna warned me that she was addicted to meth—that was one of the tools the Bandidos used to enslave their women—and would go through severe withdrawal symptoms without it. In her words, "I am going to be a real bitch for a few days." She lived up to those words.

She had no clothes, so I bought her a few things. And then I had to decide what I, a married man, was going to do with a young whore with no money and no place to go.

I decided to head south on a vacation of sorts, down to Acapulco, to see if by chance I could find some of Guerrero's legendary gold weed. We spent one night in El Paso, Texas, on our way. While we were there, I bought a few more things Anna needed. She had left the bar with nothing but the clothes she was wearing and her purse. I bought Mexican insurance on the car, a big old sedan of one type or another—maybe an Oldsmobile. I also had affidavits prepared declaring that we were American citizens.

Anna liked to dress in skimpy clothing. I told her that might not be such a good idea in Mexico, but she insisted—she said she always dressed like that and the Mexicans would just have to deal with it. She put on a pair of short shorts and a revealing halter-top. We crossed the border. As soon as we approached the customs officer manning the bridge, I knew she had made a mistake. He stared awestruck through the window at Anna's legs, not hearing a word I had to say. We parked and then walked to the building where visas were arranged. Anna was so practiced in the art of seduction that she put out those vibes without being aware of doing so—or so it seemed. A wave of confusion preceded her as she entered the building, hips swaying, invisible sparks flying, leaving men staring and their women angry. One woman hit her husband who stood frozen with his mouth open—she pushed him out of the building. Anna appeared not to notice.

At first the customs official didn't want to accept our affidavits. He also stared at Anna. I took the rejected documents from him, discreetly placed a couple of twenties between the pages, handed them back to him and asked if these new documents would work. We got visas.

Everywhere we went this wave of confusion preceded us. At one point, a bus screeched to a stop in the middle of the road, just to check Anna out as she walked around the car. We made it as far as Chihuahua City the first night.

Anna liked nothing Mexico had to offer. They didn't have her brand of cigarettes or Pepsi—at least not the Pepsi she was used to drinking. They didn't have her kind of shampoo. She didn't like the way they prepared their food. Many bathrooms had no toilet paper, a fact she found particularly infuriating. Since nothing seemed to please her in the interior, I determined to get to Acapulco as fast as possible, stopping only long enough to gas up and go on. I drove all night long, going eighty miles an hour or better straight to Mexico City. We made it through the city and continued south. About twenty-four hours after leaving Chihuahua, and maybe an hour's drive from Acapulco, we rounded a bend and ran head-on into a roadblock of Mexican soldiers. I rolled down the window just in time to hit them in the face with a cloud of marijuana smoke. The sergeant at the roadblock pointed a Thompson .45 submachinegun at my chest. I raised my arms.

My warnings to Anna were about to prove prophetic. Her body was sculpted right out of most men's fantasies, and the hot pants she had on that day hid very little. I will never forget the looks on the faces of the soldiers when she got out of that car with the cheeks of her ass showing. That ass was connected to a beautiful set of long curvaceous legs. I knew she was in trouble for sure. It didn't seem like there was anything I could do to stop it. But I knew I had to try.

I had about an ounce of weed stuffed into a tear in the seat covering, trapped deep within the foam of the seat cushion. Anna and I were ordered to the side of the car while the men searched. One of the soldiers found a couple of seeds and came running up to the sergeant with them in hand. I looked at Anna. She was scared. So was I.

The sergeant ordered our car moved to the side of the road for a more thorough search since cars were piling up behind. After about ten or more minutes, they still had not found the weed.

"Why don't you go ahead and let us go?" I asked the sergeant.

"You were smoking marijuana."

There was no sense in lying, at least about that.

"Yes, we were."

"Do you have any more?"

I lied. "No."

"We shall see."

"You've been looking for quite a while already. Come on, man, we're not smugglers. We're just down here to have a good time and play at the beach."

"Where did you get the marijuana?"

"I brought it with me."

"From the United States?"

"Yes. Let us go, sir."

"I must wait until my superior gets here."

"How long will that be?" I asked.

"Soon."

Anna and I waited for about thirty minutes, during which time I told her what had been said. She didn't understand much Spanish. She did, however, understand the looks she was getting from the Mexican soldiers. I was much more afraid of what might happen to her than anything they could do to me.

Finally a young man arrived, the equivalent of a lieutenant in our army. I could tell right away he was educated, and I suspected he had smoked a joint at some time in his life or at least had been around others who did. I decided to take my chances.

When he asked what we were doing, I told him I had come looking to buy marijuana and to have a vacation at the same time, but denied having any on us. My candid approach shocked him. He asked me how much money I had on me, and I told him $1,000. He said, "If you want to buy marijuana, then you have to buy it from me."

He told me to give him $700, which I did. But he wouldn't give me the dope himself. He told me to go inside a tent. Sitting on top of a desk inside that tent was a sack containing partially cured marijuana. The army had probably harvested the crop themselves in the last few days.

The man wanted a shot at Anna really bad, but I would not budge on that issue—she was not for sale or trade. I drove off shaking. I wasn't the only one. Anna rode huddled against the passenger side door, in a state of shock.

Once we got to Acapulco, I reached into the hole where I had stashed the weed brought from the States. It was gone. I pictured that lieutenant laughing.

* * * * *

I had a decision to make. We had barely enough money left to get home. I called Jeff Aylesworth and another guy in Fort Stockton who owed me money and told both of them to wire funds. They told me it would take a day or two at the most. Once they agreed to do it, we decided to stay. Anna liked nice things, and it didn't take long to spend the $300 we had left. In several days we were dead broke. The money didn't come for two weeks.

There was a woman who ran a villa across the bay from the major line of hotels in Acapulco. I talked her into letting us stay on credit until the money arrived. I was told it would take a day or two at the most.

There we were—in what most Americans would consider a paradise on earth—surrounded by luxury, yet trapped in our own private hell. I climbed coconut and mango trees and even dove for oysters trying to feed us, without a hell of a lot of success. I sold my pocketknife and my watch. Before it was over, I was reduced to the role of a pimp while Anna slept with several Mexican men for money—against my wishes I might add. That had to be one of the low points of my life because I was still in love with her.

I refused to eat the food she bought with the money she had earned by selling herself. Imagine that—a dope smuggler refusing "dirty money." She got drunk and I had to rescue her from a group of guys bent on having their way with her on the beach. Afterward she sobbed in my arms while explaining all the reasons she was a whore. I didn't like it, but I understood. Then I ate some of her food and shared her pain for a night. I held her in my arms and did my best to love her.

When the money finally did arrive, we headed for home—carrying my $700 sack of marijuana. Once again, our journey was not without troubles. First we got lost in Mexico City. We sat at one intersection watching the light go from red to green to yellow and back to red for over an hour without moving because the traffic going in the opposite direction always blocked the intersection when it was our turn to go. I was ready to kill someone by the time we finally moved. After four hours of fruitless effort, we decided to spend the night in Mexico City and try again another day.

The next morning, we made it out of the city. I wanted to go back by way of Durango, just to check out the area. On the way, we had a blowout in the desert of Zacatecas, in an area so remote I thought we might just die of heat stroke before we got out of there. It must have been 130 degrees on the side of the highway. I changed the tire while Anna looked on fearfully, sweating in the awful heat. I drove slow from that point forward, peering through waves and ripples coming off of the road, praying that another tire wouldn't melt and blow out. We made it to Durango, where I bought two tires and an extra wheel and then rented a room. The next morning we drove to Parral, Chihuahua, and then over to

Delicias and then north to Camargo. I drove the back road from Camargo to Ojinaga to avoid the *aduana* (customs checkpoint), and lost the muffler and several shock absorbers. We had several more flat tires on the way. Thankfully, I had more than one spare.

* * * * *

Dick Graham, my friend from the Bakersfield valley, had relocated in Ojinaga, Chihuahua, which lies just south of Presidio, Texas, on the border between the United States and Mexico. He had hooked up with a retired Mexican prostitute who had a couple of kids. Anna and I pulled up to his house, a couple of road-weary travelers. I had warned Anna about Dick, but nothing I could have said sufficed for an actual live encounter with the man. By this time in his life, Dick was a modern day *desperado.* Jesse James would have paled by comparison.

Dick—who was in the process of building a fortress in Ojinaga—had abandoned his Pentecostal wife and family, except for Tom, his only son. And nobody in his right mind was going to mess with Dick, Pablo Acosta included. Dick had built a huge cinderblock wall around his house with sharpened spikes and broken bottles along the top to prevent people from scaling it. Inside the house was an arsenal. He didn't go anywhere, not one step, without a weapon in hand. When we drove up, he came to the screen door cradling a Thompson .45 in his arms.

I introduced Dick and Anna. He introduced his new wife to us. She was a short, dark-skinned Indian woman, *pura morena,* wearing a traditional Mexican skirt. She smiled shyly in greeting as is customary with women of that type. They had just finished eating supper but invited us to sit at the kitchen table for a cup of coffee. I recounted our journey to Dick. He listened to my story and then told me he had begun to buy and sell a little marijuana, which came as quite a shock. I knew he didn't like the stuff. He told me he had shot down a helicopter full of drug agents in Mexico and described several recent repossessing exploits. I didn't know whether to believe him but sat and listened nonetheless.

He described how he believed the end of the world depicted in the Bible was near, and how the time was soon to come when cash would be abolished. And how any outlaw with a head on his shoulders should be able to see that the abolition of cash would put an end to crime as we know it. So—it was now or never. He needed cash—lots of it and quick—so he could buy a fortress somewhere and arm and supply himself for things to come. He tried to convince me to join him in several harebrained schemes to rob a bank or lure some known DEA agents into an area to buy a nonexistent load of marijuana and then rob them of their money. I laughed off most of this. Dick was dead serious.

The subject of women came up and Dick began to extol the virtues of his new wife while cursing the average white woman. She stood smiling as he bragged on her. And then to prove her worth, he told her to demonstrate her proficiency with a firearm. She produced a loaded revolver from under her apron and proceeded to take aim at the wall. Dick told her to fire. I intervened, telling her that wouldn't be necessary and convinced her not to shoot. Dick then pointed out the bullet holes in the wall where she had done previous target practice. We then watched as she ate a whole glass full of extremely hot jalapeño pepper sauce mixed with cream for desert, sweating profusely and smiling all the while.

"Yep. She'd kill a man in a heartbeat for me. Can't find women like that in the United States. And she watches my money like a hawk. I let her do all the bargaining down here," Dick said with a look of genuine admiration.

I glanced at Anna. She looked like a ghost. I don't think she said ten words the whole time we were there. Before I left, I agreed in principle to join Dick on a smuggling venture out of another region on the border where the marijuana could be bought for less than what he was paying in Ojinaga. Anna appeared thankful to get out of Dick's house alive.

I put about two ounces of marijuana in a bag and placed it in the crack of my ass right behind my balls before we started across the river into Presidio and the United States. I left the remainder with Dick after I found out he had an American border agent on the take who would allow him to cross the bridge unmolested. He agreed to bring it to me later, along with more he hoped to buy in Ojinaga.

It was dark when we arrived at the bridge. We pulled up to the customs officer with the engine roaring—the car now had no muffler. Mexican customs had applied an entry permit on the windshield when we began our journey. We were dirty. The car was dirty. There was plenty of evidence we had been in the interior of the country. So when the guy asked, I affirmed that we had been to Acapulco and had stayed for a couple of weeks. I had smoked a joint right before we crossed the bridge and figured my eyes served as evidence of that fact. The trained eye of the border agent did not fail him.

"Do you smoke marijuana?" he asked me.

"Every chance I get," I replied.

He looked surprised.

He told me to pull into the area where they did more thorough searches. We sat there waiting for about ten minutes while they ran our license number on their computer. Finally another inspector came out and asked us a series of questions and began a thorough search of the vehicle. We watched as he looked under the hood, even removing the breather cap from the carburetor. He looked under the body of the car with a long-handled mirror and then meticulously searched the interior of the car and our possessions. He banged on the gas tank and ran a probe similar to a very long oil dipstick into it. He removed the spare tire and rolled it to see if it wobbled and then bled air from each tire, smelling for the odor of drugs. The search lasted for about twenty minutes or so and revealed nothing. When the guy approached after searching the car, I concealed my nervousness. The lump of those two ounces felt like a football between my legs.

"I guess you can go—this time," he said.

"Thanks," I replied, smiling. He watched as we walked back to the car.

"You are fucking insane," Anna declared as I rolled a joint and lit it while we were pulling away from the bridge.

Early the next morning, I drove right past my house in Balmorhea and kept going, planning to leave Anna at the bus station in Pecos so she could go to Phoenix where her mom lived. A few miles short of Pecos, we had yet another flat. Neither spare contained air. A group of willing guys drove up while Anna and I were in the midst of a heated argument. I gave her what money I had left and signed over the title to the car. The men left with Anna and one of the flat tires, promising to help her get it fixed so she could get to Arizona. I started walking south. I had no money and it was forty-five miles to my house, but I didn't give a damn. We were thoroughly sick of each other by this time.

I made it home but omitted all the parts about Anna when I told the story to Cheryl. I wanted to tell her the truth and the day would come when I did. But on that day, I just couldn't get the words out of my mouth.

A couple of years later, I found Anna working in a Phoenix tit bar. And I learned a valuable lesson. Some women are whores just because they want to be.

DURANGO, THE DEVIL'S BACKBONE

Late one summer, Oscar ran out of marijuana. I loaded Crocket and Cubano into my vehicle and went to Ojinaga in search of Pablo Acosta, the notorious drug baron. Crocket told me Acosta owned the *plaza* in Ojinaga—in other words, he paid off the the *federales* for exclusive rights to move drugs through the area. Crocket said he knew Pablo. We stayed several days, waiting and talking to people. But I guess even Acosta had no marijuana at the time, and we were forced to go home empty-handed.

I drove to Plainview to collect money that Jeff Aylesworth still owed me from a previous venture. In Plainview I found out from some Mexicans who had just returned from parts further south that the new harvest season had just begun. I knew no marijuana had arrived at the border. Jeff owed me money I would never see because he had long since spent it. So I talked him into borrowing cash from several of his friends and heading to Mexico with me. His friends thought they were paying for a small quantity he could readily access. Had they known that small quantity was still in Mexico, attached to a plant, I doubt anyone would have given him a dime. As it was, we headed south in Jeff's old two-door Impala with their money, which was now my money. Jeff was a talker.

On the trip I made with Anna to Acapulco, I had attempted to establish connections in the interior of Mexico. I had found some, but I had an aversion to buying from cops. Any marijuana they had was stolen from its rightful owner. Being a farmer myself, that made those cops my enemy. I had heard the mountains between Durango and Sinaloa were traditional growing areas. I had checked it out on our way home and had liked the looks of the country, so Jeff and I headed for Durango.

The city of Durango is located on the eastern slopes of the Sierra Madre mountain range, which runs along the west coast of Mexico. It sits on a large expansive grassy plain, high in elevation, at the base of a very impressive mountain range. Numerous sawmills, large stacks of lumber and wood processing plants and factories serve as evidence of the forests above.

Once we got to Durango, we turned the Impala west and struck out for the mountains. We climbed steadily up twisting narrow roads with precarious drop-offs and sharp hairpin turns. We inched our way along the sides of forest-covered slopes, passing small towns along the way, but no major cities. The place had the right smell. Eventually we reached an area known as *el Espinazo del Diablo*—The Devil's Backbone—a narrow, natural bridge of sorts which spans two mountain peaks with steep drop-offs on both sides. The *Espinazo* is just wide enough to accommodate the two-lane highway that passes across the top. From this vantage point, we stopped and stared, surveying the country for miles to both the east and the west. The mountain fell away on both sides of this formation so sharply that we were able to throw rocks and watch them drop thousands of feet before hitting anything. Near the top grew evergreen forests similar to those found in the Colorado Rockies or the Cascades of the northwestern United States. But, unlike the deserts we had left behind, the western side supported a tropical forest some six thousand feet below—a land of bananas, palm trees, avocados, tamarinds, mangoes, and all the other plants one might expect to find in the tropics.

We began our search for marijuana from that point forward. I had bad experiences soliciting people in bars and the like in Mexico, so we decided to try a different approach. I started pulling over and asking peasants we encountered on the road point blank, "Do you know where we can find any marijuana?"

Most of these people were either on foot or on a bike so we could outrun them if they decided to turn us in. There were no phones to call ahead. All of them were shocked. Some refused to answer. Others called us crazy or something to that effect. A few pointed ahead, saying, "Further up the road." So we kept going up the road until I noticed a tiny community that really looked the part.

Jeff stopped alongside the road and let me out. The weather was cool, and water was everywhere, cascading off of the steep sides of the mountains. The air was fresh and clean-smelling with the faint odor of evergreen trees and the pleasant smell of burning wood. I walked up a slippery path to a small collection of homes. On the way I dodged a few pigs and detoured around a few vicious-looking dogs that barked and growled with serious intent. At the top of the trail I ran into a young Indian-looking woman. Right away I asked her, *"Donde puedo comprar marihuana?"*

She looked suspiciously at me and said, *"Esperese aquí."* She walked off.

I took my chances and waited as she had ordered. About ten minutes later, a thin young man—angular and dark-skinned—about my age returned. He wore *huarache* sandals, the homemade leather sandals soled with tire rubber that you see on the poor all over Mexico, and a Mexican sombrero. I saw no evidence of a weapon on him. I'll call him Daniel.

I introduced myself and repeated the question. His demeanor wasn't threatening, but he did have an inquiring, suspicious look on his face. I took that as a good sign. He didn't immediately reply but looked me over carefully. He asked me a question or two and then motioned for me to follow. Daniel led me through the tiny village and into a small opening surrounded by large boulders and lush green, tree-covered mountain slopes. An older man, a man I would later learn was Daniel's father, greeted me, and once again I popped the question. He didn't answer either, but instead began asking me a series of questions.

"Where do you live?"

"Texas. In the United States."

"Why do you want marijuana?"

"To smoke and to sell. It has great value in my country."

"Are you a policeman?"

"No."

"Do you know anyone else down here?"

"No. This is the first time I've come here."

"Why did you stop *here* looking for marijuana?"

"I heard that good marijuana is grown in these mountains. I asked people alongside the road and they directed me here."

On and on went the questions. Finally, he asked, "Do you have any money to buy marijuana?"

"Very little," I confessed. In fact, I had a couple of hundred bucks left and would need part of that to make it home.

Daniel's father made a sign to someone hiding behind one of the large boulders, who appeared with a small package of marijuana in hand—perhaps a couple of kilos. I looked it over and produced what money I could and we struck a deal. I told him I would take the marijuana back and use it to gather more money and return in a couple of weeks. He told me they had just begun to harvest but would have whatever I needed upon my return. He also made sure I understood one thing clearly. "Do not tell anyone else where you got this," he warned.

"No problem, sir."

"Also, when you come back, bring pesos. We cannot use dollars here. We will take weapons, ammunition, battery-powered tape recorders and radios, binoculars and Mexican pesos, but no dollars."

"What kind of weapons do you like?" I asked.

He produced a semiautomatic .45 from his waistband and Daniel removed a 9-millimeter from his. "We like these. And .380 automatics."

Until then, I had no idea either one of them was armed.

About the time I was to leave, a small group of armed men descended out of the surrounding rocks and forest, some with handguns, some with AR-15s, and others with *cuernos de chivos.** One at a time, they came up and shook my hand. I turned and left.

If I had been a policeman or come to rip them off, I would have been killed.

When I got back to the car, Jeff was ecstatic. We turned around and headed for home with smoke billowing out of the windows, wearing illegal smiles. Our journey back, however, was not without difficulties. Nearly all our money was gone. Before we made it back to Ojinaga, I had to trade my pocketknife for gasoline, and Jeff spent several hours trying to make a burnt set of points work by sanding them with the striker on a book of matches and adjusting the gap to various positions.

We hid the marijuana in the rear fender well of the Impala and drove over the bridge separating Ojinaga and Presidio. The inspector pulled us out of line for a search. There was no way to find the stash in that car without removing the rear seat and the internal panels that covered the hole in which we had deposited the marijuana. The seat was not removed so we rested easy, confident we would not be discovered. Maybe we were too ignorant to know we should have been afraid. After the search we headed for Balmorhea and then Plainview.

All kinds of people were willing to finance our next trip once we had mari-

* The horns of a goat, derived from the curved shape of the clip. Also known as AK-47s.

juana to show them. There was no marijuana anywhere near Plainview or Lubbock, at any price.

A new season had begun.

* * * * *

Jeff and I headed back to the mountains of Durango for another load. This time we bought about fifteen pounds or so, smuggled it over the bridge, sold it in Plainview and returned to the mountains for more. By the third trip the load filled the stash compartment. The Impala would hold only thirty pounds in the fender wells without being so tightly packed that it would emit a dull thud when a customs officer tapped on it from the outside of the vehicle. So now we faced a new dilemma. We had outgrown our stashmobile.

On the next trip we filled the trunk with suitcases containing marijuana. I knew from the trip I had made with Anna how to circumvent the Mexican customs checkpoint by taking the bad road from Camargo to Ojinaga, but the route was hell on cars. We made it anyway, minus the muffler and after fixing a flat or two.

Jeff and I didn't like the idea of crossing the bridge with marijuana in the trunk, so we drove to a remote area on the Mexican side near Ojinaga, packed what we could get into the stash compartment and hid the rest in the desert. We drove across the bridge, got searched, but survived. We hid that weed on the American side in an arroyo using branches, brush and cacti to make it look natural, then crossed back into Mexico for another load. We crossed the bridge three times that day to get all the marijuana across. We got searched all three times. We had pushed our luck as far as it could be pushed and were going to have to come up with a better plan. The stress we endured that day was just too much to bear.

But after we sold the marijuana, the abundant cash in our pockets and all that it bought made the stress seem a little more bearable.

INSANITY

Word—and evidence—of marijuana of better quality than my friends produced in the mountains of Durango arrived at the river, and consequently in Balmorhea and Fort Stockton, too: a lime-green *sin semilla*[*] produced in large government-sanctioned fields in the desert climate of Northern Mexico.[†] Our customers wanted the best their money could buy, so we decided to make a run to Santa Elena. Jeff came across someone in Plainview in jeopardy of losing a fancy black Camaro Z-28 to a creditor. The car was insured against theft and the owner owed more against the car than it was worth. Jeff drove me to a house and handed me a set of keys. The owner told Jeff he would not report the car stolen until eight the following morning. Jeff gave him a pound of marijuana for the car, without the title.

---

[*] *Sin semilla* refers to female plants which have not been pollinated. The flowers remain intact rather than falling off to produce seed, and the clusters of flowers, more commonly known as buds, grow much larger than they would if pollinated. The resinous female flowers are the most aromatic and potent part of the plant.

[†] Powerful forces within the Mexican government had sanctioned and were protecting these fields of marijuana for obvious reasons. A single field was so large, rumors say, that up to 3,000 laborers would be working a crop. The marijuana produced in these fields was roughly five to seven times stronger than traditional Mexican marijuana. Not only was this product much better, there was much more of it, and its emergence began to challenge the products of other countries which had dominated the market. These fields were unknown to the United States until discovered by DEA agent Enrique Camarena. He was assassinated in 1985.

I took the keys, opened the door, started the car and drove off. Jeff followed along behind in my Ford Thunderbird, a 1969 model with a 429 engine that should have been illegal. The power that engine generated was more than the chassis and suspensions of that day could safely contain. It was midnight. We had eight hours to get the Camaro out of the country. I drove south to Lubbock. Once we got on more remote roads, I opened up the throttle and drove a hundred miles an hour for an hour. Then we switched cars and drove faster.

Daybreak found us parked on the American side of the river crossing opposite Santa Elena, Chihuahua, in the Big Bend National Park. The river was low, maybe two feet deep, but the bottom was strewn with large rocks and the clearance on the Camaro was minimal. I figured maybe we could drag it through the river using horses or a tractor.

Santa Elena* was a sleepy little town just south of Castalón, which was nothing more than a small store and a ranger station on the American side of the river. It was very hard to get to from the Mexican side. There was a forbidding mountain range to the south and terrible, rutted rocky roads connecting it to the rest of Mexico so locals routinely crossed the river to buy things from the American side. The streets were dirt: dogs, goats, chickens, hogs, horses and burros all roamed aimlessly around the town and lounged under large cottonwoods which provided shade. The residents of Santa Elena had a few fields that they irrigated out of the river, but no significant production came from them. Some of the people ran a few cattle, but the main source of income was the business of smuggling drugs into the United States. Many of Santa Elena's residents had family members who lived in the nearest American towns of the Trans-Pecos region of Texas. These people distributed their products further north.

A group of young men generally hung out near the river and made money rowing tourists across the stream which separated the United States and Mexico. They also sold small amounts of weed to those who came looking for something to smoke. It was foolish to show up in Santa Elena with a pocketful of cash. Stories are told of those who went there and were never again seen. But if you knew the right people, houses full of marijuana were available, and no Mexican lawman did anything to stop the business. *Federales* never arrived unannounced. When they did come, the people were forewarned and there was nothing illegal to be found.

I waded the stream separating the two countries and approached the old

---

* Post-9/11 Santa Elena is almost dead.

man who I had met through my farmhand José Chávez after my initial trip. He was the principal dealer in town. We made a deal. I didn't get much money for the Camaro since I had no title—about fifteen hundred dollars in credit, which translated into about six or seven pounds. Jeff had given the owner of the car a pound, but the five or so pounds we cleared were worth at least four thousand bucks in Plainview. I had cash to buy more to complete the load.

"How can we get the car across the river?" I asked.

"I'll take care of it," the old man replied.

The old man sent his son, a guy about my own age, across the river in a boat. While I watched from the Mexican side, he took the keys from Jeff, opened the hood to the car and removed the fan belt so the fan wouldn't throw water on top of the engine. Then he got into the car, started it, and—to my utter amazement—drove it down the embankment and into the river. How that engine kept running with the top of the hood under water and the exhaust bubbling from below the surface remains a mystery to me, but it did. I kept waiting for the current to sweep him away or a rock to stop the car's progress or the engine to die, but none of this happened. He made it all the way across the river.

I crossed the river later that evening on foot with a tow sack full of marijuana. We secured it in the old T-bird and headed north.

⋆ ⋆ ⋆ ⋆ ⋆

Jeff was adept at stroking pool sticks and women, and both of these skills created problems for him and me. We walked out of more than one bar with money that had walked into the bar in someone else's pocket.

Jeff was not a fighter. While I didn't go around looking for fights, fights had a way of finding me when I was with Jeff. I was so thick-headed, I didn't even realize he used me this way until one day when we were in Plainview. A man about 6'3'' and weighing close to three hundred pounds walked up to our table and began to beat the shit out of Jeff. Jeff just balled up in a defenseless posture. I forced my way between the two and received a shot to the face for my effort. Before the man could hit me a second time, I moved in, grabbed him and took him to the floor. We grappled. I ended up on top of the man with a stranglehold on his neck while the rest of the people in the bar tried to pull me off of him. Eventually I let him up but not before I choked him nearly unconscious. I ended up with a big shiner out of the deal.

Later I found out that Jeff had slept with this man's daughter and had given her some dope. Jeff had also beaten the man out of quite a bit of money at pool.

Another time, Jeff and I went down into the black section of Plainview—what the white people called "Colored Town." White men rarely went there. Excepting us, that is. We went to an illegal club which didn't even open its doors until all the rest of the bars had closed. Cars surrounded the place for two blocks in every direction. Inside was wall-to-wall black. The place was so crowded that everyone damn near had to move at the same time in order to dance. The bar served two kinds of beer—Coors and Budweiser.

Jeff led me through the crowd to a booth. He had a pound of marijuana in his hand. Several large black men were sitting there. They looked over the weed and bought it on the spot. We watched while they rolled the entire pound into tiny, thin joints, most of which were sold before we got out of the place.

"How can a place like this operate?" I asked Jeff.

He answered with a question of his own. "If you were a cop, would you come in here?"

I looked around. "No, I don't guess so—not without an army, anyway."

A day or two later, we went back to Colored Town to sell a second pound of weed. This time we went during the day. Jeff walked into the bar, which was deserted at that hour, without even trying to conceal the large Ziploc bag of marijuana. Once again two men inspected the weed. Jeff quoted them a price. They thought the price was too high.

The men produced money and said, "This is all we'll give you for it."

"No, sir," Jeff replied. "I'm not coming off of the price."

The men exchanged looks, and then one said, "Well then, I guess we'll just take it from you."

Jeff shrugged his shoulders. In a flash, he pressed a small semi-automatic pistol into the head of one of the men.

"No, sir, I think I'll take your money instead," he said.

I started the car while Jeff came running—with both their money and my weed in hand. I don't know about Jeff, but that was the last time I went to Colored Town in Plainview. I never took part in another robbery, but I can't say I feel guilty about relieving those fellows of their money.

* * * * *

As time went by, the waits I had to endure to collect money from Jeff became unbearable. The money I spent waiting severely cut into potential profits I had coming, especially when my whore habit was figured into the equation.

I gradually began to shift a larger and larger portion of my business to my cousin Phil. Besides, since I had kidnapped Anna, there was nowhere for me to party safely in Lubbock. The Bandidos had put a contract on my head, or so I had been told. And the whore I now loved was named Michelle and lived in Fort Worth. Phil participated with me on occasion in the smuggling portion of the job, but I think it fair to say that smuggling was my specialty—his was selling the stuff. We made an odd pair.

Phil was extremely intelligent but got bored easily and hadn't figured out a good way to turn any of that intelligence into a profit. He was slightly built and an admitted homosexual. Phil smoked Camel cigarettes non-stop and was nervous in any environment. When smuggling, he literally shook.

One time he decided to come down to La Pantera crossing with me to pick up a load. I had converted another car similar to Jeff Aylesworth's Impala into a smuggling car, with seven-by-fifteen-inch six-ply truck tires. We drove this car down to the river, bought a load from Oscar who had brought it up from Piedritas and used horses to cross the river into the United States. Then we concealed the load in the stash compartment and waited through the long night. The following morning, right before daylight, we struck out, headed north.

Phil had smoked at least ten packs of cigarettes going in and coming out of the park in less than a two-day period. There were times when he forgot he already had one burning and lit a second. When this happened, he alternated between the two, puffing on one and then the other, interrupted only by an occasional hit on a joint I handed him.

I rounded a blind corner of the dirt road leading out of the Big Bend National Park driving way too fast. This road was referred to as River Road on a map. It was just above the cutoff that led to the San Vicente crossing. I looked up just in time to see the heavy-duty grill of an oncoming pickup truck. I hit the brakes and braced for impact. We collided head-on. Both of us were trying to stop so the speed at impact was not great and no one was hurt, but my car's front end was crushed. The fan was driven into the engine. Battery acid leaked out onto the ground.

A Mexican national drove the pickup. He had three passengers crammed into the cab with him. We got out to survey the damage. His truck wasn't damaged and started right up. My car was going nowhere.

"Should we wait for the rangers?" he asked.

"Do what you like," I told him, "but there's something you should know."

"What's that?" he asked.

"That car is loaded with marijuana."

The man's eyes doubled in size. I had never before seen this guy, but I figured he, like everyone else in the area, knew what went on around there and was probably not a threat. He asked for permission to leave.

"Of course," I told him. "If you don't mind, send word to Oscar Cabello about what happened here," I added. "Do you know him?"

"Yes, sir, I will."

The men got in their truck and continued on, leaving us sitting in my broken-down car.

Phil wanted to get the marijuana out of the car and hide it in the desert. I insisted we leave it where it was. It wouldn't be easy to find the weed without disassembling the vehicle. I had gone so far as to superglue the screws and bolts holding the inner panels in place; they would have to be chiseled off.

We sat and waited. About nine a.m., a park ranger headed in the opposite direction pulled up.

"What happened here?" he asked.

"We had a head-on collision."

"What happened to the other vehicle?"

"They went on."

"Who were they?"

"Hell if I know. A pickup load of Mexicans."

"Where did they go?"

"I don't know. South."

"Probably went to San Vicente crossing. I'll see if I can catch up to them."

The ranger drove on down the road. About thirty minutes later, he came back.

"I couldn't find them. I guess they already crossed the river." He looked at us and asked, "Where were you guys going?"

"Fort Stockton."

The ranger asked a few more questions and then offered to call a wrecker. We promptly accepted. Phil and I rode all the way to Fort Stockton in a wrecker owned by the United States government with a driver from Ireland. On the back of the wrecker rode a totaled car packed with marijuana.

AT THE MERCY OF THE DESERT

I changed vehicles constantly, never spending much money on one because I knew I would have to get rid of it once I'd been seen in it. Sometimes I bought a decent car and this worked out well and other times I bought a junker. More than one ended up abandoned on the side of the road with a blown engine, just miles from where I purchased it. One old truck I bought, however, captured my heart. I couldn't bear to part company with the thing.

It was a white 1966 International pickup. I bought it off a lot in Fort Worth. I built a camper shell with a false bottom, installed it in the back and smuggled several loads. I did more work on the vehicle and then, fearing I might be recognized, I drove into the parking lot of a motel—two cases of spray paint and several sore fingers later, I drove out in a brown pickup.

I had been to Piedritas with Oscar from the crossing at La Pantera. I even had followed him through the maze of roads leading there but had never tried to get there on my own. Nevertheless, the day came when I needed to. Oscar was already there, waiting on me. I was reasonably certain I could find my way.

I got the money together and went over the list of things I carried: a fully loaded tool box, a jack, several spare tires, a lug wrench, a five-gallon can of water, a five-gallon can of gasoline, a logging chain, matches, epoxy glue, a shovel, tire tools and patches, a come-along, an air compressor that plugged into a cigarette lighter, and a small air tank. I never went anywhere without the knife I carried in my pocket, another in a scabbard on my belt, rolling papers, a cigarette lighter and marijuana to smoke. I also bought a couple cases of Coors beer to take to my friends. Then I struck out for La Pantera.

There was no town at La Pantera—better known among gringos as Talley Crossing—and to my knowledge there never has been. The crossing occupied a piece of land at the southernmost tip of the bend in the *Rio Bravo del Norte,* aka the Rio Grande, right before the river dumps into Mariscal Canyon. The untrained eye would never suspect there is a crossing there. But a closer look reveals that the banks are accessible from both sides of the river, though not directly across from each other. To cross from the American side, a driver has to enter the water and drive upstream in an arc, along a shelf of sorts, and come out some fifty or so yards upstream from where he had entered. I once saw what could happen if someone drove straight into the deep channel directly across from the entrance on the American side when an empty school bus tried to cross that way and was swept away by the current.

The one fault the old International had was a tendency to die when exposed to water. On a previous crossing near Lajitas, I found myself in the middle of the river trying to dry out the points with a can of ether to get it started again. The pickup was loaded. This time, however, crossing the river proved uneventful.

While the roads on the American side of the crossing may seem rough and the country wild, they pale by comparison to what's on the Mexican side. South of La Pantera there is no road—or what Americans would call a road. There are, however, paths which have never seen the blade of a bulldozer. These were formed when someone drove arbitrarily through the country in the general direction of where they wanted to go. And then someone else came along and copied that previous effort. Over time, floods eroded these paths in the many arroyos they crossed, leaving treacherous, gaping holes ready to swallow a vehicle whole, denying passage. Considerate travelers sometimes repaired these holes with shovels, branches, rocks and dirt, but—more often than not—they just backed up and looked for an easier way to get around the obstacle. Over time, this created a honeycomb of paths, most of which were traps, ready to snare the uneducated driver foolish enough to be there.

The best route to Piedritas from La Pantera took about three or four hours to navigate—a horse could easily outdistance a truck over the terrain. Straying a couple of inches in the wrong direction could mean falling into a deep hole or turning a vehicle over. Thorns and sharp rocks were sure to cause flats. When they did, you had to deal with the flat yourself because it was a long way to help. No water was available for miles. Summertime temperatures were brutally hot and the nights surprisingly cold. Every plant or animal you came across seemed armed to deliver pain—the plants had thorns, and the animals, fangs or stingers. People—even Mexicans familiar with the area—died trying to drive through that country.

I followed the trail as best I could, always choosing the heaviest traveled path at the many forks I came to. Then I arrived at one place where recent traffic seemed to be equally divided: one path went left, one path went right. I couldn't remember this place. I chose the path to the right. A couple of hours later, I realized I had made a mistake.

I had no idea where I was going, but I did know I was going through territory I had never seen before. And then, as if that weren't bad enough, a thorn pierced one of the veins in the pickup's radiator and the engine began to overheat.

I added water to the radiator and watched as it poured out through the tiny hole. I took the cap off the radiator so it wouldn't build up too much pressure, crimped the metal vein shut as best I could with a pair of pliers, and continued on. Another hour went by and I had to add more water. By this time, I was getting scared. Finally, I poured the last of the water into the radiator and had to add the five-gallon can of gas to the tank as well. I wondered if I should turn around but decided I had come too far. The path had been traveled so it had to come out somewhere. When the engine heated up again, I stopped in despair. *Was I going to die in some godforsaken part of the Mexican desert?*

But then I had an epiphany of sorts—*When in Mexico, think like a Mexican.*

I saw a mesquite bush alongside the trail covered in thick thorns. I removed one of the thorns and jammed it into the hole. It fit snuggly and stopped the leak. Having no more water, I began to open and pour beers into the radiator. I had enough to fill it. A few hours later I arrived at La Morita. I knew to take a left to go to San Miguel and then another left to get back to Piedritas. By the time I got to Piedritas, the gas was gone and all the beer I had started out with was in the radiator. But I was alive and among friends. They all got a laugh but mourned the loss of the beer.

The pickup smelled like a brewery from then on.

* * * * *

Not long after this, I needed to deliver some money to Oscar. My Levi jacket, my pants—everything was stuffed with money—probably $15,000 to $20,000. We arranged to meet at La Pantera crossing. I was not going to bring a load back on this occasion, so I took Cheryl along for a vacation of sorts. Four of our children rode in the makeshift camper built on the back of the old International pickup. Cheryl carried the baby, Windy—our fifth—in the front seat. Willie, a black Labrador retriever, was also riding in the camper.

Rain began to fall—not hard but steady—as soon as we turned onto the river road. Ominous looking clouds engulfed the peaks of the Chisos mountain range to our west. Flashes of lightning and the roar of thunder filled the sky. It was obviously raining much harder up there than where we happened to be.

I approached what looked like a narrow but fast-moving body of water coming down an arroyo, stopped, checked it out and decided we could make it through. The front tires of the old International eased into the water and we started across. The water was only about a foot deep, but the old truck's engine died halfway across. A little moisture had found its way to the distributor cap.

In a matter of seconds, the water began to rise in the arroyo. I shouted at Cheryl and told her to get out and head for the bank with Windy. By the time she got the door open, the water had risen a foot. I jumped out of the cab and went to the back of the truck, grabbed two of my kids and headed for the bank, which was no more than ten or twelve feet away. By the time I got back to the truck, the water had reached the steering wheel. I grabbed another child, but Lisé, my oldest daughter, was scared and would not come to the back of the pickup. I yelled at her and she finally came. With two children clutching tightly to me, I forced my way ashore a final time.

Willie, our black Lab, watched from the back of the truck, but by now it was too late. The truck was almost totally submerged by rushing, brown, boiling, muddy water. Seconds later, we watched as the truck was carried off by a huge wave of water. Willie popped to the surface. He tried to swim toward us as we called him, but he was carried away. Several hundred yards downstream he managed to get out and return to us.

We sat waterlogged on top of a nearby hill and watched the flooding arroyo. An hour later, the stream of water had stopped and nothing but a few puddles and twisted debris remained as a sign of its ever having been there. A man in a four-wheel drive Jeep came along and gave us a ride back to the Park Ranger headquarters. The Rangers were kind enough to take us to the Chisos Mountain Basin where I rented a cabin for a few days. Luckily the money I carried was in the pockets of a jacket I wore. We spent the next few days hiking around the basin and had a great time.

I lost my pickup. It was found buried in a sand bar, sticking straight up, about a half mile from where we tried to cross the arroyo. A few days later my cousin Phil came and picked us up to go home. I never tried to retrieve the old International. It was too well known by then and would never have been the same. I got bills and notices from the Park Service but ignored them, figuring they could sell what was left to pay the bill.

I mourned the loss of the pickup, but not much—things could have been much worse.

PIEDRITAS

Once I figured out how to get to Piedritas on my own, I began to do it routinely. This allowed me to vary the time of my return in case someone tried to follow me down to the river. It also gave me the opportunity to spend time in the village getting to know the people and the countryside.

Piedritas lies between two mountain ranges that run in a north-south string. The only way a traveler would encounter it is if he or she specifically wanted to go there—the road entering Piedritas from the south is the only road connecting it to the rest of Mexico and it dead-ends there. Another twenty-five miles as a crow flies separates the town from the southern tip of the Big Bend National Park and the border with the United States. Twelve miles to the south an east-west road passes through a small town called San Miguel. This is where you can find the cutoff to Piedritas, but no sign indicates its presence. Very few people, even in Mexico, realize it exists.

Since Piedritas was an *ejido,* the land was actually owned by the Mexican government, but the right to use it was shared by the *ejiditarios.* In theory, an *ejido* is a communal farm and ranch operation. However, the reality seldom resembles the theory. The land belonging to the *ejido* at Piedritas was big, perhaps a thousand square miles of territory stretching from a point about seven miles south of the village all the way to the Rio Grande. Members of the *ejido* had grazing rights to all of this land. They also had rights to use an irrigated field under a gravity flow water system and a reservoir of water that collected in a lake about six hundred acres in size and at least fifty feet deep near the dam.

At that time, a hundred households and six hundred residents shared the town. Of that number, however, very few actually owned any livestock. The system of *ejidos* was designed to do away with the huge haciendas owned and controlled by wealthy families. Yet such efforts rarely panned out. Someone had to rise to the top and run things or nothing got done. Then the whole damned thing ground to a halt.

Those without any animals had to find some way of feeding themselves. Most found themselves working for someone that did have livestock and/or equipment. The only other business options available to the poor of Piedritas were digging fluorite or collecting *candelilla* wax, neither of which paid more than five dollars a day. Both occupations required tools and/or livestock.

Because Oscar had money coming in from his smuggling business, he had more cattle and horses than all the rest of the *ejiditarios* combined, so most people worked for him. He was the only man who could afford to buy farm equipment, fencing materials or major expense items the town needed. And he was the only one capable of getting his voice heard in places where policies affecting the *ejido* were made.

A large portion of the day in Piedritas was spent taking care of the basic necessities of life. Water was pulled from a well in the middle of town using ropes with buckets attached. It was poured into a variety of receptacles, some on wheels, some not, some powered by people, others pulled by burros and still others in the back of pickup trucks. People gathered firewood and sold it. Or cleaned houses and washed clothes by hand.

Most of the houses had dirt floors. In the early mornings, the town's women wet these down and swept out the dirt inside and in front of their homes. No one had electricity, but some did have refrigerators which ran on propane or kerosene. Kerosene lanterns or propane lamps were used to light the homes.

Chickens, goats and hogs provided most of the meat eaten in the town. The men harvested field crops by very primitive methods. They cut wheat and oats by hand and bundled the stalks into *manojos*, or shocks, and then later separated the grain from the straw by running over piles of *manojos* with the wheels of a pickup. Afterward they tossed the crushed material into the air and allowed the breeze to carry away the straw. Corn was picked by hand and stored on the cob.

One way people passed the time was to sit around stripping the dry kernels from the cobs, using the end of one cob to strip another. This corn fed not only the people but also the burros and horses.

The pace of life in the town was slow but steady. There were few handouts available so whoever didn't get up and do something might just starve. So everybody did something.

On several occasions, I climbed a hill comprised of odd-shaped boulders rising out of the desert floor, stacked in unusual, seemingly precarious positions. I found dim traces of ancient paintings on those rocks made by people long since dead. In the evenings, just after dark, these same rocks served as a meeting place for many of the young men of the town to gather and talk between puffs on crudely fashioned joints. Unlike Oscar and the older generation, a fair percentage of the young men smoked marijuana.

I listened while these men described their lives. Many wanted to go to the United States to work. They asked lots of questions. They teased and joked among themselves, just like kids anywhere do. I noticed that youngsters of the opposite sex rarely, if ever, hung out together during daylight hours. They could do so only in certain situations like dances held in a small courtyard. Celerino provided a generator used to illuminate the yard and to power a stereo system for these dances. I never participated, but I watched while they drank and danced. The young men and women did find time to slip off and do what young people do anywhere, especially after their parents got good and sloshed. There were also a number of married women whose husbands had gone off to work in the United States. When I listened closely from those rocks late at night, I heard stifled giggles and moans—when *Sancho* came around to visit.*

I also discovered that the young people in Piedritas weren't allowed the option of participating in the marijuana trade. Only one other man other than Oscar and his crew did, a cousin of his by the name of Macario Villarreal. Macario was a brutal man who operated on a much smaller scale than Oscar. He hated gringos and I was no exception. I suppose because he was related to Oscar, he was allowed to stick around. Stories were told of Macario going across the river and stealing the possessions of American tourists in Big Bend National Park, including their vehicles. I have no way of verifying these stories, but he sure seemed the part—a modern day Mexican *bandido*.

---

* Among Mexicans living in the United States there is a common joke. "I'm sending money home to Sancho," they say, referring to the men who stay behind in Mexico and secretly service their wives as Sancho.

* * * * *

One day a traveling preacher showed up. The guy preached hell and brimstone and summarily accused, indicted and damned the town to hell in a loud booming voice. I have little tolerance for such types, so I wandered away from town to sit on the rock pile. Even from there I could hear the guy rant and rave, hour after hour. Finally the time came when reconciliation was offered the miserable sinners, to hear him tell it, directly from the throne of God. All they had to do was to repent—and then, of course, contribute to the cause. It chapped my ass to see the guy taking money from people so poor their basic needs weren't being met.

I thought I had escaped the preacher. Later that evening I showed up for supper at Celerino's house to discover we were to share the meal. I guess since Celerino had built the church and was known as a Christian it was his duty to receive him. The meal was served and the preacher began to recount his good deeds of the day and the powerful message he had delivered. At one point he asked me what I thought of his performance.

I told him I had left the town it was so hard for me to bear.

He looked at me, almost in a state of shock.

"Why do you say that?" he asked.

"You come here and take money from people so poor they have no food. You should be collecting money somewhere else and bringing it here to give away. And you would probably damn someone like me to hell because I sell marijuana to help them out."

Celerino and Oscar stared with mouths agape and Oscar's mom excused herself from the kitchen. I kept eating.

"Get away from me, Satan," he finally stated.

"I'm not Satan and I'm not going anywhere," I told the guy.

"In the name of Jesus, I command Satan to leave this room!" he shouted.

I think both Oscar and Celerino wanted to melt by this time.

"I told you, fellow, I'm not Satan and I'm not going anywhere."

He jumped up and ran out of the house. Looking back, it appears Satan listened after all. Both Oscar and Celerino apologized on his behalf but somehow made amends with the preacher as well. The next day he sat in the streets but preached no more. I am told he went crazy shortly thereafter.

* * * * *

One day, an old man with long white hair and a long white beard rode into Piedritas on a burro. Another burro trailed behind, packed with merchandise. No one ever told me who he was or what he did to survive, but I felt certain I had seen a holy man. His skin was weathered and brown, but his eyes and face shined and an aura of goodwill emanated from him. I saw him only twice and each time his presence moved me. I never said a single word to the man—or he to me—except to say hello.

I climbed a mountain near Piedritas and near its top found a wooden cross. I considered the cross, and the reverence people have for this symbol and then another thought occurred to me. *Why have we made holy the instrument of torture used to kill the one righteous man ever to inhabit this planet?* To me it is akin to worshipping a gun or an electric chair.

AMADO CARRILLO, THE LORD OF THE SKIES

My first and only face-to-face encounter with Amado Carrillo, the notorious Mexican narcotraficante, took place in Torreón, Coahuila, in 1984. At the time, my cousin Phil and I were moving a couple of hundred pounds every two weeks or so from northern Mexico to the Dallas-Fort Worth area.

I disliked distributing and selling marijuana and felt like my "niche" was the smuggling end of the business. Oscar Cabello was at that time my principal source in Northern Mexico and provided me with a consistently high quality *sin semilla*. The Mexican product produced in these government-sanctioned fields compared favorably with much more expensive homegrown varieties raised and sold on the West Coast and was superior to the Colombian seed weed going around at the time. I took a load of his product to the West Coast and got positive feedback. My friend James and his associates in Oregon were capable of moving massive amounts of marijuana, and Oscar told me his guy—who turned out to be the infamous Amado Carrillo—had access to fields of the stuff and thousands of workers. I talked to Oscar about arranging to put his "big guy" with my "big guy" to see if we could get in the middle and make some serious money without the hassles involved in selling and collecting on relatively small quantities. But first Oscar arranged that Phil and I would meet his "big guy."

Phil and I drove to Torreón, Coahuila, and met Oscar at a motel called *La Posada*. I called my friends from the West Coast and we waited for Oscar's man . . . and waited . . . and waited. I had always operated on the premise that a moving target was the hardest to hit and didn't like the idea of sitting in one place too long. After a couple of days in the motel, I became uneasy and expressed my concerns to Oscar. He assured me all was well and introduced me to a large group of men at a bar in the motel. "After all, we're buying from the cops!" he said laughing, as they tossed down another round. I assumed the men I saw were, in fact, policemen. They all carried .45s on their hips. They were loud, obnoxious and drunk, but—in spite of all the backslapping and smiling—I was acutely uncomfortable.

Finally Amado showed up. He, Oscar, Phil and I met in our motel room and talked for four or five hours between lines of cocaine provided by Amado. Amado was a slender but unremarkable young man dressed in cowboy boots and Western clothing. I remember thinking he looked very young to be the *mero chingón* I had heard he was.

During the meeting, I remember Pablo Acosta's name coming up. I told Amado I did not want to work with people like Acosta, who had a reputation for killing people, and that I wanted only to make money and move herb without having to harm anyone. I told Amado that my connections wouldn't come until I had verified that a load was ready and that the quality was good. Then, and only then, would they come.

He and Oscar left around midnight, leaving behind a small amount of coke for Phil and me. We immediately disposed of the rest of the coke in vacuumlike fashion and went to bed. Around three a.m. there was a loud knock on our door. Phil was closest to the door and got up to open it. I sat up but remained in my bed.

The door flew open. A man with a machine gun pointed squarely at my chest stood there. I saw fear, intensity, anger, frayed nerves and hatred in his eyes—those eyes and the barrel of that gun were all I saw. It's amazing how knowing you are about die can focus your attention. A large group of men followed him into the room, all armed and yelling loudly, issuing orders for us to get up and get out of the room. We were forced into the parking area—naked. When I tried to stop and pick up my jeans on the way out, one of the men jabbed me in the ribs with the point of a rifle barrel. That's the last time I ever went to bed naked.

Our assailants all wore clothes similar to those Amado had worn—Mexican cowboy dress: boots with steep riding heels, Western-cut slacks, and button-down Western shirts—except for their apparent leader, who might very easily have been mistaken for a lawyer. He wore a guayabera and expensive trousers made from high-quality thread and Italian loafers with a fresh shine. This man interrogated both Phil and me at length and repeatedly asked, "Where is the dope? Where is the load?"

"What dope?" I asked in reply. "We don't have any dope."

"Why are you down here?"

I decided to level with the guy. "We came down to buy marijuana but we haven't been able to get any. We don't have anything. Look all you want. There is no dope!"

The guy didn't want to accept this. Finally, remembering what Oscar had said about our suppliers being cops and running the show down there, I asked hesitantly, "Have you talked with Oscar?"

"What Oscar?" the well-dressed man asked.

"You know . . . Oscar!"

He motioned to a couple of his men and the four of us walked toward Oscar's room. By this time, the men had mercifully given me a blanket to wrap around myself. Being naked was cold as well as embarrassing that night. The door to Oscar's room was open when we arrived.

"Is this the Oscar you asked about?" the man asked.

Oscar was laying face down on the floor. A man stood over him, Uzi in hand. Oscar weighed around two hundred eighty pounds or so—the man standing on his back was all of a hundred and thirty. The small man repeatedly kicked Oscar in the ribs with the heel of his riding boots—over and over again—in the same spot, until a lump the size of a grapefruit protruded from his side, red and blue and ugly. Oscar grunted each time the bastard kicked him but said nothing. The others stood armed and ready, looking on while the small man worked.

Finally, they pulled Oscar into a sitting position. His face was bruised and swollen and blood streamed from his nostrils. My heart sank when I looked into his eyes. Oscar was a close friend as well as my business associate. I felt totally powerless.

The men then led me back to my pickup where I was made to remove a door panel to expose a gun we had secured there. Phil had told them about it. We were interrogated further and then sent back to our room, which by this time had been searched. The door slammed shut behind us. We waited, expecting the worst. An hour or two went by. The sun came up. We peeked through the curtains and discovered that things seemed to be back to normal outside. Around eight or nine, I gathered my courage, opened the door and walked out to a world apparently oblivious to what had happened just hours before.

I walked to Oscar's room, knocked and waited. He answered after a couple of minutes. He was barely able to walk and grimaced with pain at the slightest movement.

"What the hell happened? I thought we were buying from the cops."

"They were *judiciales* from Nuevo Leon. As soon as they were able to get through to Amado on the phone, a payment was arranged and they were called off."

"Bullshit, Oscar! You said we were buying from the cops. If that's the way they treat their friends, I want nothing to do with these guys. I don't know about you, but I'm getting the hell out of here. Are you coming?" I asked.

"Yeah, I guess so."

Phil and I loaded Oscar in my pickup and drove north to Piedritas. Oscar winced at every bump in the road and rode in a hunched-over position. I was afraid he was bleeding internally and might die. We made it to Piedritas and left Oscar at his father's house. It took him weeks to recover from the savage beating.

I never accepted Oscar's explanation of what happened that night. I was convinced Amado sent these men to test us, to see if we were cops or informants.* I guess we passed the test. While I never again met Amado Carrillo face-to-face, I continued to do business with him through Oscar and was eventually successful in getting some loads to my guys on the West Coast, although never on the scale I had envisioned. That night did little to develop the confidence of my guys in Oregon, and they damned sure weren't going to Torreón to meet anyone after hearing about what had happened to us.

I learned that, contrary to accepted history,† Amado Carrillo supplied Pablo Acosta and just about anyone else moving dope through that part of the world. He was in fact already becoming, as the leader of the Juárez Cartel, the *mero chingón de los chingones!* It would be a number of years before the American DEA would discover that fact. When I escaped from prison and lived as a fugitive in northern Mexico, I would see for myself the scale he was operating on.

Years later, reading Terrence Poppa's *Drug Lord* and Charles Bowden's *Down by the River*, I was astonished to discover how violent Amado Carrillo became. I won't ever forget that I badmouthed Pablo Acosta for his way of doing business in Amado's presence. What a young fool I was!

---

* Oscar disagrees. He now says the men were from Interpol and that even the *federales* were at risk of arrest until Amado intervened. Oscar also told me he had given a sample of heroin to an informant, something neither Phil nor I knew anything about.

† Terrance Poppa says that Carrillo replaced Pablo Acosta but, in truth, he was the leader from the beginning.

DESPERATE TIMES

The business of making and keeping money by smuggling marijuana is much rougher than most people know. Everyone seemed out to get me. In dealings with the big-time border suppliers, most loads were at least partially sold on credit, and there were very few good excuses for not paying. Occasionally Mexicans sold me "shit weed" that no one wanted because better was available and I'd get hung with the stuff, laying around in a motel for days, weeks or even months, spending money trying to sell it. Weights were not always as advertised. It was no fun to go back and argue over such matters. The fact remained: if someone ripped me off or got busted, I still had to pay.

Cops were after me. Rival dealers, rip-off artists and thieves—even whores looking for valuable information to sell to cops—became potential enemies. The people I sold to could easily turn on me if they got busted or ripped off. They could become greedy and take my money to buy some other guy's dope rather than pay what they owed.

The business was made worse because marijuana is a perishable commodity. It must be grown, cured, packed and preserved in the proper fashion to be marketable. Packed too wet, it will mold. Packed too dry, it will fragment and become "shake." The U.S. consumer is hard to please. Under the conditions in which it's grown and processed, it's often difficult to come up with a consistently good product, but of course, that's what the people want and demand. Although marijuana is a seasonal product like tomatoes and chili peppers, consumers expect it to be there, day in and day out, throughout the year. When you don't provide, they go elsewhere with their money—or worse yet, with yours.

At times I was rolling in money. There were other times when something went wrong or my sources dried up and I was left penniless. I often had to go to extreme lengths to get seed money to reestablish my business after a drought.

Which leads me to another story, or a number of them I guess.

* * * * *

Late one summer I had been unable to find anything other than leftover shit no one would pay for. The time was right for Daniel's first few plants to be nearing maturity. If my hunch was right, I would be the first to arrive home with a fresh product. I was so broke that I had to borrow $500 from a potential buyer in exchange for a special deal when I got back.

I paid $140 for a one-way airplane ticket to Mazatlán, where I slipped through customs without any luggage and then rented a ten-dollar-per-night motel room.

The next day, I turned my dollars into pesos. None of the growers I knew in the mountains would accept American money for marijuana because spending or exchanging dollars in their community would create suspicion. I walked to an outdoor marketplace, choked my way through the flies and stench and bought a large canvas backpack and a poncho for the mountains. I then went to the bus station and bought a ticket to Durango even though I would be getting off in the mountains halfway there.

At that time there were several classes of buses operating in Mexico. I chose the cheapest available—one of those that went down the road belching black smoke. All the seats were occupied and people stood in the aisles. Nevertheless, the bus driver stopped for anyone and everyone wanting a ride, cramming them in for a little extra cash that went right into his pocket.

I shared the bus with squealing pigs sacked in burlap bags, live chickens trapped in the tight grasp of their owners' fingers and the odor of sweaty unwashed bodies basting in tropical heat. By car the drive usually took four hours, but due to plenty of stops and the speed of the bus as it labored overloaded up the grades, it took us much longer. When we got to Daniel's village high in the mountains, I wove through people to the front of the bus and asked the driver to stop. The other passengers stared. There were very few reasons why a young gringo would want off in these remote mountains.

I watched the bus pull away and waited until it rounded several bends before climbing a slippery trail. The path led to a collection of crude houses constructed from rough-sawed lumber. Daniel's wife recognized me and walked up dressed in a traditional long skirt favored by women of the area, greeted me with a handshake and a smile, and then went in search of her husband. No men were visible in the village, but I knew they were there—watching. Always watching.

Mazatlán had been hot and humid. Up here, it was cold, so I put on the poncho. I smelled the crisp mountain air laced with the wonderful smell of burning firewood.

Daniel came walking up dressed in the clothes of a peasant, though not the traditional variety one would expect of a man older than he. Light green jeans and an embroidered leather belt covered his lower body, and a battered hat rested upon his head—the smile of familiar eyes accompanied his handshake. His well-worn huaraches soled with tire treads protected his rough, calloused feet and a quick glance revealed the lump of a 9mm handgun I knew would be in his waistband. Every male member of this community from twelve on up was armed.

Knowing it was illegal for men to carry arms in Mexico, I had once asked Daniel what they did if cops showed up.

"We shoot them," he replied.

"What if the soldiers come?" I asked.

"That depends on how many come. If a few come, we kill them. If too many come, we run and hide."

"What if they catch you?"

"Then most likely they will kill us."

"Can't you pay them off?"

"No, not us. It is said that others can and do. We are poor. We cannot do this."

At this particular village, the men never slept in their homes for fear of being caught asleep and being arrested, opting instead for the safety of the mountains with no cover other than a sleeping bag and a sheet of plastic to protect them from the elements. It rained damn near every day in those mountains. At that elevation, it got cold. During the day, they would go to their homes but not without setting up surveillance on the road. There was no way short of a helicopter to approach the village without being seen far in advance of your arrival.

I had only enough money to buy five pounds of marijuana at the equivalent of fifty dollars a pound if I were to keep enough money to get home. I explained this to Daniel. He had just begun to harvest early buds, but he had enough decent *sin semilla* bud to fill my backpack—around ten or twelve pounds. He knew that if I were successful, I would make a huge profit and be in a position to buy far more the next time.

He asked, "Can you sell *goma*?"

"What?" I asked.

"Goma . . . opio." He showed me some dark gummy crap wrapped in a piece of plastic.

"No," I replied, and then a stupid thought came to mind. I disliked heroin and wanted nothing to do with buying, selling or using it. But I did have acquaintances—heroin addicts—who used the stuff. It occurred to me they might be able to smoke opium to ease their cravings while breaking their addiction to heroin, so I bought the small piece of raw opium he offered me.

To avoid drawing attention to his village, he guided me several miles through dim mountain trails to another spot above the highway where we waited for a bus to come along. When one approached, I said my good-bye, ran down the side of the hill and flagged it down.

I boarded the bus with my backpack full of marijuana. The bus's passengers greeted me with curious stares. I paid the driver to go to Durango. He smiled knowingly. There were no unoccupied seats so I stood near the front, right behind him. Religious artifacts adorned the front of the bus and I couldn't help but whisper a prayer as we departed. I'd almost swear the driver joined me in that prayer. I sweated through the tiny town of La Ciudad, where I knew soldiers often checked traffic but on this day none were there. We arrived at a bus station in Durango crawling with cops. I got out like I owned the place, walked right past the cops up to the ticket counter and bought a ticket to Camargo, Chihuahua.

Once again, I boarded a bus but this one was a Chihuahuense bus with two daredevil drivers and a souped-up engine. The drive to Parral and then over to Camargo was a white-knuckle ride all the way. By the time we arrived, I felt lucky not to be represented by one of those white crosses along the road.

The bus continued on to Chihuahua, but I knew if I went that way I would have to go through a checkpoint between Chihuahua City and Ojinaga and that was something I was unwilling to try. At that time the road from Camargo to Ojinaga was incomplete but paved for ten miles on either end. The rest was dirt, rocks, terrible potholes and washouts. The *aduana* checkpoint was rarely manned.

I spent the night in Camargo. The next morning I strapped on the backpack and began walking toward Ojinaga. After about five miles, I got a ride with road construction workers in the back of their dump truck. They let me out where the pavement ended and I began to walk again through the hostile desert terrain. This time I covered about ten miles before a pickup load of men stopped to give me another ride. I hoisted the backpack into the bed of the truck and climbed in behind it. The pickup continued on, bumping merrily over the washboard dirt road.

A few miles later we came across an old man with a huge sack of plastic bowls and dishes, bundled in a tarp. The driver of the pickup stopped. The man muscled his wares into the bed of the truck and joined me.

The man was heavyset and had on the clothes of a peasant but seemed to be content with his life as a traveling salesman. I could tell he was curious about what a gringo might be doing in this particular part of the world. We talked. For some reason, I decided I should tell him what I was up to before we arrived at the tiny village where the *aduana* was located.

"You have what?" he asked incredulously when I told him about my cargo.

"Marijuana," I replied.

With a look bordering panic, he replied, "You should not do that here! That is very dangerous. Have you not heard of *El Víbora*?"

"No."

"*El Víbora* controls that around here. If you want to buy marijuana, you must go see him. He will kill you for what you do! You must talk to *El Víbora* to do that here!"

I had never heard of *El Víbora*,* but this guy sure as hell had and the stories had left him terrified. He didn't want to be anywhere near my sack full of marijuana. He started to tell me all about Carrasco and the danger of operating in his territory. He insisted that doing that was a good way to end up dead. Years later, I read about Manuel Carrasco in Poppa's *Drug Lord* and learned that at the time he was no longer considered the main man around Ojinaga. Pablo Acosta had taken his place. Apparently this plasticware vendor hadn't heard that yet. Before we got to the village, he knocked on the back window of the pickup truck, ordered them to stop, unloaded his bundle in the middle of the desert and continued walking with the heavy load, shaking his head in disbelief as we drove off toward the checkpoint.

I myself was tired of walking and decided to take my chances, hoping that

* Manuel Carrasco, the guy who owned the plaza before Pablo Acosta.

no *federales* would be manning the station. I watched apprehensively as we approached and then breathed a measured sigh of relief when we found the station unoccupied. My ride carried me all the way to Ojinaga. I checked into another cheap motel and stashed my backpack in a room. That afternoon, I walked through the streets of Ojinaga until I spied a group of young street thugs.

"Do any of you know how to get across the river on foot?" I asked them.

"Where do you want to go?" one small man asked. He stared at me with one eye. The other eye strayed off to the side, distracting me.

"Marfa," I replied. "I'll pay someone to take me."

"Why don't you walk across the bridge?"

I took another big chance.

"Because I have marijuana."

"How much will you pay me?" he asked.

I offered him fifty dollars to act as my guide, but I told him he had to come with me right then and there or not at all. Past experience had taught me well. This guy wasn't going to leave my sight until we crossed the river.

We went back to my room, retrieved the backpack and rented a taxi to take us to a small community about fifteen miles upriver. I bought a kilo of Gamesa crackers (similar to shortbread cookies), and then we took off—like so many wetbacks on any given night. We walked under cover of darkness and then, for no apparent reason, we stopped. He signaled me to be quiet.

"Why are we waiting?" I asked.

"They are over there . . . *la migra!"*

Sure enough, after about fifteen minutes, lights came on. The Border Patrol was searching for us. Rather than continue north, we walked downstream and waited again. We finally reached an area where he thought it would be safe to cross. To my surprise and confusion, there was no river to cross. Apparently the Rio Conchos, a tributary that enters from the Mexican side farther downstream, supplies most of the water I was accustomed to seeing around Presidio and through the Big Bend region. All the water upstream from that location had been diverted for irrigation purposes. What was left was neither *grande* nor *bravo*. There was, however, a man-made ditch, about ten feet across and waist deep in mud. My guide removed his pants and shoes and waded through, suggesting I should do the same. Foolishly, I didn't remove mine. I ended up with about twenty pounds of sticky mud on my Wranglers and two shoes full of mud. On the other side of the ditch, we encountered a large strip of plowed ground the border patrol maintained to tell where people have come across. We walked backward across this area, hoping to fool them into thinking we had crossed heading south.

We played cat and mouse with *la migra* the rest of the night, walking for thirty miles over rough, rocky, irregular terrain. I kept telling my companion we were going in the wrong direction, but he insisted he knew where we were. We drank from puddles in the bottom of arroyos and filled discarded plastic soda bottles for later. I was thankful not to be able to see the water because I had to drink something. I almost fell asleep walking. At daybreak, we stopped.

We watched as the first light of a new day began to emerge. He told me he wasn't going any farther. I'd be on my own from that point forward. During our conversation, he described several arrests and the torture he had endured in Mexico. To prove his point, he pulled up his pants and showed me horrendous scars where Mexican cops had wrapped bare electrical wires from an extension cord around the calves of both legs while his feet rested in a bucket of water. Then the cops plugged the wire into a wall socket. He also told me that his eye strayed because they had repeatedly shocked him in the face with a cattle prod, causing him to lose control of the travel of that eye. The damage was permanent. I could see the leftover fear and psychological damage in his face as he described these events.

I paid him the agreed-upon price. Before he left, he told me that what I was doing was foolish. Then he took a loaded .45 from his waistband and said, "If I were a bad man, I could take all your money and the marijuana too."

"You already have all of my money!" I replied, smiling, opening my empty billfold to prove the point.

There was nothing left for him to steal other than a backpack full of marijuana in hostile territory. I'm fairly certain that was his original plan, but somewhere along the line he changed his mind.

"¡Estás loco!" he said, and then turned to head back to Mexico. *"¡Vaya con Dios!"* he added, over his shoulder. Sometimes there is a bond between crazy people.

I watched as he headed back toward the river. I was so exhausted, I could barely keep my eyes open. I lay down and went to sleep on hard-packed ground littered with rocks and pebbles and slept like I was on a Serta. When the sun rose higher into the sky, the heat followed. I woke up sweating and took a look around by the light of day.

To my dismay I was only about two miles from Presidio, which meant that—including the walking I had done the day before between Camargo and Ojinaga—I had covered about forty-five miles on foot during the previous twenty-four hours and was only two miles into U.S. territory. My clothes were caked in mud and my feet were wet and blistered. I was still sixty miles from Marfa.

I got up and started trudging through the desert. I had planned to walk all the way to Marfa. I made it about fifteen miles and couldn't go any farther. I was bone tired. The blisters began to bleed and each step hurt terribly. I was so exhausted that my body almost shut down on me.

In desperation, I walked out to the highway and began flagging cars. When a Border Patrol vehicle came by, I flagged him too, figuring that was one way to ensure he didn't stop. He drove on by. Finally, a well-dressed Mexican national who worked for the Mexican consulate picked me up. I must have been a hell of a sight. I placed my backpack in the trunk of his sedan and climbed in, issuing apologies for soiling his seat. He dropped me off on the streets of Marfa.

I started walking again, this time toward Fort Davis. I got a ride with another Mexican national in a pickup truck. I guess I looked too dirty for anyone from my own country to stop. I tossed the backpack into the bed of his pickup, and we headed for home. He delivered me at the doorstop of my house in Balmorhea. I was able to sell the marijuana for $10,000 and return to Daniel and his village for more, but I will never forget the price I paid for that load—a price that can't be measured in dollars. It took several days for me to recover so I could walk again. I gave the opium to my *tecato* friends after smoking a little to see what it was like.

"You can smoke this to ease your cravings," I told them. In less than a day, they had figured out some way to cook the shit up and shoot it into their veins.

So much for good intentions.

BUSTED ON THE BEACH

A larger house with a few irrigated acres came up for rent in Balmorhea. A man named Scripps owned the house, the son of *the* Mr. Scripps of Scripps-Howard publishing fame. The house had six or seven bedrooms but was plain, constructed from cinder blocks with a huge basement and a pool table. Previous renters had left the place in a shambles. I made a deal to clean it up and make repairs in exchange for rent. For a time, Scripps got his money's worth.

Mr. Scripps was an eccentric fellow who had married a Mexican national. Between them they had raised twelve children in Balmorhea. He was a genius with multiple interests. The property housed a foundry—complete with a large furnace for heating metals and equipment to handle and reconfigure the molten products—and elaborate bee-keeping and honey-extraction equipment.

Mr. Scripps wore khaki clothing, drove a flatbed pickup with a standard transmission and no air conditioner, and had more money than God. If you saw him, you'd never guess this to be the case. Locals tell stories of helicopters descending upon the town to whisk him away to New York to attend board meetings of powerful publishing corporations. After his children were raised, he and his wife moved to Fredericksburg, Texas.

I seldom saw him.

Next door to the Scripps's rent house lived an older white fellow we referred to as Happy. Happy was from other parts, maybe California, and appeared to be the product of a previous generation of rebels like Ken Kesey and the Merry Pranksters. He and his wife looked like hippies. She had an almost unnatural affection for a boa constrictor snake. Both smoked marijuana, and it was not uncommon to find them in some sort of meditative state, complete with burning incense and the lingering odor of pot smoke. If you disturbed Happy in this serene state, you might get your ass shot off.

I don't remember exactly how the subject of marijuana came up with Happy, but it did. Sometimes we shared a joint. A friend of his named Danny came to live with them. He had been severely burned in a natural gas fire when he was working a construction job. His partner on the job was killed. Danny received third-degree burns over significant parts of his body—the worst of which were on his face and hands. In the process of dealing with the pain caused by this condition, he developed an addiction to opiates. When the doctor stopped supplying them, he turned to heroin.

The insurance company settled with him—$30,000 was the price he received for his disfigurement. Knowing $30,000 wouldn't support his habit for long, he began to look for ways to invest his money. Through the grapevine, he learned what I did and he approached me with a plan—if I would take him and a part of his money to Mexico, he would pay for a load of marijuana, cover all expenses, help me smuggle the load and give me half of it upon our return. I agreed to do this.

We struck out together, headed for Mazatlán. He wanted to make the trip a vacation as well as a business venture. Previous experience had taught me the two didn't mix well, but I agreed anyway. On our way to Mazatlán, we stopped by Daniel's place high in the mountains, bought a kilo of marijuana and made arrangements to pick up a large load on our return trip.

Danny was doing without heroin. To deal with the withdrawal symptoms, he made me stop at nearly every pharmacy we passed where he bought an assortment of pills and bottles of cough syrup containing codeine which he drank like most people drink beer. To be honest, I didn't really like this fellow and found his habits and compulsions annoying, but business is business and can't always be done with people you like.

We rented a motel room away from the beach and spent several days changing dollars into pesos. Danny wanted to hang out at the beach so he could check out the beautiful women. So that's what we did, eating fine seafood and drinking Don Pedro brandy with Coca-Cola and a squeeze of lime.

Mazatlán is home to some of the largest, best tasting shrimp in the world. I understand they're called prawns when they attain that size. I enjoyed the ones grilled over natural charcoal and drizzled in butter best of all.

One night, after stuffing ourselves with good food, Danny decided he wanted to go check out the beach. The kilo of marijuana we had bought was in the luggage in the trunk of our car. We drove to a secluded area near a rocky portion of the coastline, rolled a few joints, parked the car and picked our way down a steep cliff. We sat on a large rock and stared out into the ocean, with nothing but the moon and stars to illuminate the night—smoking joints and watching and listening as the waves smashed into the shore below. The breeze hit us directly in the face as it blew inland from the sea.

This same breeze carried the smoke off the end of our joints up the cliff and into the nostrils of two soldiers assigned to guard the presidential palace located above us. Of course, we didn't know this.

Our meditation was interrupted by a shout from above. I turned to find myself staring into the end of an Uzi held by a guy dressed in the green fatigues of a soldier. I tossed the joint—lot of good that did.

The soldier ordered us to climb up the face of the cliff and then escorted us at gunpoint to the top where another soldier waited.

The soldiers questioned us and then forced us to open the car. Both of us watched in horror as one of the men found the kilo and turned toward us, holding it up with a grin.

A familiar sinking feeling hit me, but this was no time for grief. I knew our only chance was to deal with these guys. If it went higher up the ladder, the prospect of working things out would be much more difficult and costly. Danny spoke no Spanish and our captors no English, so it was up to me. We were told that their relief and their immediate supervisor would be by in an hour or so. We would be passed off to him when he arrived. From there we'd go to jail.

While we waited, I began to feel them out. At first, neither of the men seemed receptive to letting us go.

"How much money do you earn a month?" I asked one of them.

He told me he earned the equivalent of two hundred and fifty dollars a month.

"I have more than that in my billfold. It won't do me much good in jail," I suggested. "And someone is going to let me go for it. If not you, then the judge."

"You Americans think you can come down here and do as you like. Like your money can buy anything. You are in *my* country now. And I am going to teach you a lesson," one of them shouted.

"I mean no disrespect to your country," I replied. "We're just down here to have a good time. And maybe to make a little money."

"When I went to your country to work and also to have a good time, I was caught by your *migra.* They beat me and tortured me. I will never forget that. And those men had better hope and pray I never see their faces here in my country. Now *you* come here and enjoy the best Mexico has to offer, things *I* can't afford in my own land. You are in Mexico. *This is my country, pinche gringo!"*

"Take it easy, man. Do you think I'm a friend of the United States government? They throw people like me in jail. Please don't hold us responsible for the actions of my government."

Initially, it didn't seem like these men would release us. We were led to a bus stop and forced to sit on a curb, supposedly waiting for their superior who would be by to take us to jail. After a while, though, the two men began to talk back and forth. One wanted to let us go—the other didn't.

Finally they came back to see how much money we'd be willing to pay. We had about five or six hundred dollars on us and gladly gave all of it to them.

PSYCHO

I wanted to bring in larger loads overland—the part of the business with which I was most comfortable—but to do so was scary. Marijuana is bulky. Bringing loads over roads without being able to conceal the merchandise was asking for trouble. You never knew when you might encounter a roadblock. When I was discussing this dilemma with a redheaded, dope-smoking cowboy friend of mine by the name of Bird, I suggested I might be interested in trying to bring in loads using burros. Bird lived in Pecos but said he knew a ranch hand who lived and worked twenty miles south of Marathon. This man knew the country well enough to show me around. He would also be able to provide a storage facility for the merchandise and a place to load vehicles once the way was clear.

On our way to see this guy, Bird warned me that he was *not all there.* I figured that was okay. That put us in similar company. Nobody smuggling marijuana was *all there.*

We arrived at this fellow's ranch after dark. Bird called out toward the house while I waited in his pickup. The door opened. A man stood in the light, cradling a wicked looking mini-14 rifle in the crook of his arm. Bird talked with the guy for a moment and waved me in.

Like always, I had a supply of marijuana. Like always, Bird had a supply of ice-cold beer. We entered the small ranch house packing our goods. The man I met wore glasses and appeared to be a common everyday redneck. The only thing I found unusual about him was that he never put down the rifle. Where he went, the rifle went too.

We began to smoke and drink and then to drink and smoke. In between hits, I laid out my proposal. I sat on a couch near the front door. Our host sat opposite me on a kitchen chair and Bird sat to my right in a recliner. My plan was to walk all the way from the border with a pack train of burros by night carrying maybe a ton of marijuana. I would hide out during the day. My thought wasn't an original one. I had met Mexicans who told me of the days when they did the same, loaded with *candelilla* wax. I had also met others who made forays into the United States to steal cattle and drive them home. If this were possible, bringing marijuana by the same method would also work.

I knew such a feat would be difficult—exceedingly difficult—but that's why it appealed to me. Most smugglers tried to beat the cops at their own game—using modern devices and technology—and found themselves always one step behind. Cops had unlimited resources—after all, their employer prints the money. I was counting on the cops being lazy and unwilling to go out and sweat or to go anywhere their vehicles couldn't carry them. I suspected they would think it improbable anyone else would want to do that either.

Our host also thought the plan might work. He had Mexican ranch hands who routinely walked from the river to the ranch, and he personally knew where most of the gates were and where we could find water, cover and feed.

He also understood my logic. Illegal aliens rarely came through the area, except for those specifically walking to a particular ranch—and there weren't many of them. Ninety miles of desolate country separated this guy's ranch from the border, with no real town in between. The country was unforgiving. The Mexican side was even more desolate with nothing but dirt trails leading to the border. Mexicans didn't even go there. Go far enough east and you hit Del Rio and the Mexican town of Ciudad Acuña. Go far enough west and you hit Presidio and Ojinaga. The area in between is known as the Big Bend, one of the most remote places on the North American continent. The Mexican side is referred to as the *despoblado*—uninhabited land or *badlands.* All American cops had to do was sit back and watch three roads—the only roads leading out of the region. Nobody in his or her right mind would try any other way of getting out of the Big Bend overland.

As the beer and marijuana took effect on our host, I began to notice strange behavior. I caught him peering at me through thick glasses with an uneasy smile on his acne-scarred face. I found the look less than friendly. He sat with the mini-14 across his lap. When he went to the refrigerator to get a beer, he carried the rifle, and returned to the same spot. I noticed the fondness he had for this weapon and asked him about it.

Bird began to nag the guy, saying, "Tell him about your rifle."

It was, he said, in essence the same as an AR-15, except the stock was made of wood and the barrel was chromed. He told me about the muzzle velocity, the trajectory and the velocity at any given distance, the killing range, how the bullets would tumble when hitting the ground in front of the target and possibly jump up and do damage in spite of being fired too low. Both Bird and this fellow were very drunk at this point. I never did drink much but had smoked plenty of marijuana. I always smoked plenty of marijuana during this phase of my life. The more to be had, the more I smoked, especially if there was plenty to be had.

And I was good at ensuring there was.

"Tell him what we used to do, down by the river."

"We'd shoot targets," he said, with delight dancing in his eyes. By this time, the guy was beginning to stroke the rifle, almost as though he was masturbating. I began to worry.

"Tell him how I got this scar," Bird said, holding up his arm to reveal a scar about the thickness of a small water hose.

"I burned him with the barrel," he told me.

"Do what?" I asked.

"He shot the rifle until the barrel was red-hot and then branded me on the arm with it," Bird said, laughing hysterically. His friend joined in on the laughing.

I stared in horror.

"Tell him what else you shot," Bird said.

"I shot Mexicans," he said proudly. He didn't get the reaction he had hoped for.

I will sit and smile and try to be sociable, often agreeing with things I don't like or believe, just trying to get along with people. This, however, went against the grain of my very being.

"You're shitting me, aren't you?" I asked.

"No," he replied, with a serious look.

"You sorry fucking bastard. What the fuck is wrong with you?" I asked.

No sooner had the words escaped my lips than I knew I had made a mistake. It was as though I had slapped him in the face. Bird continued to goad the guy into giving more details but the guy clammed up and looked steadily at me.

His face drained of blood. He began to tremble slightly. Bird continued to laugh and joke about what they had done. His friend wasn't laughing. He sat quietly, stroking the stock and barrel of the gun—and staring. At me.

I measured the distance to the door, trying to remember the layout in front of the house. I tried to visualize how far I would have to run before getting out of sight. I had one thing going for me—the night was pitch-black. By this time, it was midnight. I waited for the guy to get up and get another beer, but when he did, I saw there was no way I would have time to get out of there without getting shot. He kept an eye on me all the way to the refrigerator and back.

Finally, I decided to confront him with the thought pounding at my brain.

"You want to shoot me, don't you?" I asked, knowing the answer.

It startled him. I repeated the question.

"Yes," he finally responded, breathing, almost panting, in shallow breaths.

"Why?" I asked.

"I don't know. I just do," he responded.

I spent the next five minutes trying to talk him out of this insanity. Bird thought it funny. I was petrified. If there are such things as demons in this world, he had at least one working on him. I don't think Bird really had any idea how close this guy was to shooting me.

It must have taken another hour for me to convince this guy to let us go. He did so only after I promised I wouldn't tell anyone his dirty little secret. Then I had to persuade Bird to leave. He was shit-faced drunk and perfectly content right where he was. He saw no reason to go anywhere. After a number of attempts, I finally convinced him to take me home.

The crazy bastard watched from the door of the ranch house as we drove off. He was illuminated in a halo of light, rifle in hand, staring out into the darkness. I experienced an overwhelming wave of relief. I cussed Bird bitterly once we got to the highway and were out of rifle range. I came close to beating the shit out of him for having put me into such a predicament, but he was too drunk to know any better or even care. So I let him be.

I regret to say that I kept my promise to this guy for a number of years. In fact, I only recently told this story to a few people. I don't know if the events this crazy bastard described actually even happened. But I'll bet they did.

Needless to say, I never did get around to going back and doing any deals with the guy. From that day forward I got an eerie feeling whenever I had to pass by the ranch where he worked.

Even now, I fear the consequences of telling this story. I fear the consequences of not telling it more. If what the guy said is true, he needs to be dealt with.

I sometimes wonder how many more like him are out there and how many dead Mexicans lay buried out in the desert whose only crime was a desire to come North to work for an honest wage.

THE ORGANIZATION

I got a call from my cousin Phil one day, telling me that his connections Jeff Tyree and Kenny Schulbach wanted to invest in our business. Jeff often bought large quantities of marijuana from Phil for resale. He apparently did the same with Schulbach, but for drugs other than marijuana. However, we had the better connections for marijuana so Schulbach wanted to join our operation. I told Phil to bring the whole crew out to Balmorhea. From there, we'd go to the border for a load.

Jeff Tyree was 6'3'' tall, with long blonde hair and a lean tanned body not unlike the body of the wrestler Gorgeous George in his prime. He wore a ring on every finger and gold chains around his neck. He didn't use drugs but sold just about any flavor the public demanded. He wore a beeper and would get up at any time of the day or night to service his customers. He often rented a limo and a driver to haul him around and changed women more often than I changed socks. Aside from dealing dope, he acted as an agent for various aspiring musicians in the Dallas/Fort Worth area.

Kenny Schulbach was a well-groomed, tall, slender man, older than the rest of us—perhaps in his mid- to late thirties and a seasoned veteran of the drug culture. He abstained from using drugs too. In fact, he wouldn't even drink anything that contained alcohol or caffeine.

Kenny still had connections to members of the hippie movement from the Haight-Ashbury district of San Francisco. He hoped to get something going in Fort Worth and in California as well. He was intelligent and knew everything about the latest gadgets and devices used by members of both sides of the drug war—things like night-vision equipment and scanners which would pick up law-enforcement frequencies. He was a gun freak with access to automatic weapons, grenades, anti-personnel mines and who knows what else. He also had a lot of cash. Unlike most dope dealers, Kenny watched his money closely and reinvested in his business rather than blowing it. I liked the idea of going in with Kenny because he would connect us to larger buyers and eliminate exposure to the street dealers Phil had to deal with.

Jeff, Kenny and Phil arrived in a white Chevy van. We took another vehicle along so Jeff and Kenny wouldn't have to ride out in the loaded vehicle when we came back. We drove to La Pantera, crossed the river and met with Oscar on the Mexican side. Oscar had some very good high-quality *sin semilla* that day, more of the kind grown in the large government-sanctioned fields, and we bought around a hundred pounds for $250 a pound.

For the first hour of our meeting, Kenny walked around with a grenade in hand. Anybody who shot him would soon be dodging shrapnel. I guess it would be fair to say Kenny was not the trusting kind of fellow—particularly in the company of a group of unknown Mexicans on their turf.

We repackaged and sealed the marijuana in long thin rolls of Seal-a-Meal plastic. We had brought along a device that melted the two layers together at either end, creating an air- and smell-proof package. We were careful not to touch the outside of the bags any more than necessary to prevent transferring the smell of the marijuana from our hands to the packages. We then washed the long thin packages in a strong disinfectant-type detergent and placed them into the body of the van after we removed all the interior panels. We even lined the packages with fabric softener sheets—the kind added to a dryer full of clothes to make them smell nice—so they would get by dogs trained to sniff out drugs. Then we reinstalled the panels and camped out overnight. The next morning, we drove off. Jeff, Kenny and Phil sold the marijuana for $1,000 a pound. Our $25,000 had become $100,000.

We repeated this process over and again but with more vehicles and progressively larger loads. After several trips, neither Jeff nor Kenny bothered going back to the river, leaving that end of things to me.

★ ★ ★ ★ ★

Kenny and Jeff did a couple of trips with us down to Durango too, because the profit margins were so much greater when we bought the weed directly from the growers instead of purchasing it at the border from Oscar.

By this time I had bought the white van from Kenny with proceeds earned from our joint business venture. Kenny had customized the van into a dope mobile. It had huge shock absorbers on each front wheel and would run a hundred miles an hour at half throttle. Large 727-landing lights rode on the front grill and a special alternator generated power to handle the drains those lights demanded. Phil also bought and equipped a maroon van with a large stash box; we traveled in tandem.

At night I drove in front on desolate stretches of back highways, using those powerful lights which made day out of darkness for well over a mile in front of the vehicle. I turned them off when I saw an approaching vehicle—looking into these lights would render oncoming traffic blind. Once or twice I blinked my regular lights in order to get someone to dim theirs. When they didn't, I hit them with a short blast of those landing lights, almost sending them into a ditch.

We also began to take weapons to these men in Durango. In Mexico this can be more serious than being caught with marijuana. We survived one search at a checkpoint when the Mexican cop had us pegged but was just too lazy to tear the vehicle apart to find what he suspected. And then again, maybe the *mordida* had something to do with his being lazy. This guy told us he liked cream with his coffee, so we made sure he had plenty of money to buy both.

BLOOD ON THE PECOS

After Dick Graham had once again gone broke farming, he sold his farm in the Bakersfield Valley to a Mexican national from Ojinaga. He also reverted back to repossessing. But now the targets of his efforts were often people he had done business with around the McCamey area, who had—for some reason or other known only to Dick—fallen out of favor.

A D-8 bulldozer was doing some work near the farm co-op. One night, Dick started it, loaded it on a semi trailer, hauled it to an area near Presidio, drove it through the Rio Grande into Mexico and sold the damned thing! The next day found him back near McCamey as though nothing had happened. This was only one of what I am sure were many repossessions Dick made during that era of his life. A few weeks later he showed up at my house in Balmorhea.

Dick had tried to talk me into joining him in various crimes, among them a REAL bank robbery involving taking ALL the cash in the bank by hostile takeover. Another time he tried to convince me to join him in a reverse sting operation on some known DEA agents—to sell them a large nonexistent load of weed and then forcefully take their money at gunpoint when they arrived for the buy. I tried to convince him of the folly of such ideas, but he was hellbent on doing something really bad in a really big way. After listening to several more of these ludicrous schemes, I talked to him about the virtues of the marijuana business and the great sums of money that could be made without having to risk killing or being killed if all went well.

"Can you sell the damned stuff, though?" he wanted to know.

"If it's good, I can."

I told him I had connections in Plainview and a cousin in Fort Worth who could move large quantities of marijuana. I also told him about my Mexican connection in Santa Elena. Finally he agreed to do a load with me—as partners. Dick would put up money and/or merchandise to buy the load. We would smuggle the marijuana together. I would take it to Plainview to sell and split the net proceeds with him. His son Tom would also assist in the smuggling and sell a small amount near McCamey. Dick wanted to meet my connection in Santa Elena before doing the deal. One evening, we headed for the river. On the way, Dick talked about his philosophy, which for him meant a life of constant crime.

"I am not going to be arrested. No one will ever put a set of handcuffs on me. The only way I'm going down is shooting and the only way they'll take me in is feet first!"

I tried to convince Dick that it was not necessary to kill or be killed to make money in the marijuana trade. After all, I reasoned, the worst that could happen to a guy that gets busted is a year or two in jail—wasn't it? Once again he reaffirmed his position.

"I'm not doing one day in anybody's jail."

To make his point, Dick told me about one time when he was in jail. It was in his hometown and he was young. He had witnessed jailers abusing a young kid in their custody. They had made this kid duck-walk and quack while they ridiculed him. I was sure the young man he was describing was him. Dick then pulled a chrome-plated semiautomatic handgun out of his waistband and said, "No sir, I'm not going to jail."

I had heard Dick talk tough before and had just figured that's all there was to it—talk. I remember one time a year before when he fronted a pound of weed to my friend Leroy Hernández in Balmorhea telling him, "I'm leaving this weed with you. When I come back—either you have my weed or the money. If I show up and you don't, I'm going to cut off your ear and add it to my collection."

I figured this also was just more bullshit. Knowing what I know now, I wouldn't be surprised to discover there are a few guys walking around without an ear, courtesy of Dick Graham. In any event, Dick was convincing and tended to get paid for his dope.

We arrived at Castalón at sundown and got a spot in the campground a mile or so upstream from the crossing. Large cottonwoods provided shade in the campground. We did our best to rest under them for a couple of hours. As soon as it was good and dark we walked down a path to the river and waded across to Santa Elena.

A man I had previously done business with greeted us at his home—not the main man in town, but a worker of his whom I trusted. He was short and dark-skinned with a large smile and calloused hands with a strong grip acquired by plenty of hard labor. He invited us into his house. We talked over a cup of coffee in the kitchen. I translated since neither spoke the language of the other. Dick wanted to know how much the weed would cost and what kind of merchandise the Mexican might take in exchange. The man told him he needed tires for his pickup and that guns and ammunition always had value in Mexico. We looked at a sample of the marijuana he had available and agreed to return at a designated time to complete the deal. Dick told the man he would bring four tires. We then waded the river and walked back to the pickup, formulating our plan on the way. Dick wanted to head home.

I didn't like to leave the campground at night but, since we had nothing to hide, I decided it would be okay. We started driving north around midnight. Fifteen miles north of Castalón, a pickup with flashing lights pulled in behind us. It was a park ranger. Dick stuffed the chrome-plated handgun into the waist of his jeans, pulled his shirt out to cover it, pulled the pickup to the side of the road and stopped.

The ranger opened his driver's side door, positioned himself behind it and directed a spotlight at us, gun drawn and pointed in our direction. He identified himself and ordered us out of our vehicle. We stepped out and into the strong light behind Dick's truck, me from the passenger side and Dick from the driver's side.

"Have you been to Mexico?" the ranger asked, still behind the protective cover of his door.

There we stood, still dripping river water from the waist down.

"Yes," I replied before Dick had a chance to lie.

"What were you doing down there?" he asked.

"We just went to visit a friend of mine," I replied.

"Did you bring anything back with you?"

"No."

The ranger holstered his weapon and walked toward us.

"Do you have IDs?" He asked.

We handed him our driver's licenses. He returned to his vehicle and called in our license numbers over his radio while we waited. A couple of minutes later, he returned.

"You didn't bring anything back with you?" he asked again.

"No, sir," I replied again.

"Do you mind if I take a look in your vehicle?" he asked.

"Go right ahead!" Dick replied. While we stood behind the pickup, the ranger entered Dick's pickup from the driver's side and searched for contraband. While still in the cab of the truck, he hollered out, "Are there any weapons in the vehicle?"

"No," we both replied in unison.

The park ranger walked back to our location.

"Do either of you have a firearm on your person?" he asked.

"No," I replied.

The ranger looked at Dick.

"How about you?"

Dick squared off at the man and stared. He didn't answer.

"Do you have a firearm on you?" the ranger asked again.

Dick just stared, smiling now, and said nothing. I disliked what I saw. The policeman was facing me. Dick was in front of me and slightly to one side. If they shot at each other, I was liable to get hit by the ranger's fire.

I looked straight into the ranger's eyes. He was tall and wore glasses. I saw the blood drain from his face. Dick had both hands on his hips like a gunslinger ready to draw, and he wasn't going to answer the cop's question. The ranger didn't know what to do. Seconds ticked by. Finally, he made the choice that saved his life and possibly mine.

"I guess you can go then," he said. He handed each of us our driver's licenses. His hands shook as he did. The ranger got into his truck, turned around and headed back toward the river.

I couldn't believe it. We got back into our vehicle and continued on.

"Holy shit!" I said. "I thought you were about to shoot that guy."

"I was. The only way he was going to see my gun was when I pulled it out to shoot his ass. Like I told you, I ain't going to jail."

"But we didn't do anything illegal, Dick."

"No matter. I'm not giving my gun to anybody. And I am not getting arrested."

We drove on. I began to have doubts about doing any more business with Dick Graham.

"We're still on, aren't we?" he asked.

"I guess so, but I'm going to show you how to do this without having to hurt anybody. I just want to sneak some dope into the country, sell it to people willing to pay cash money for the stuff and go on about my business. This doesn't have to be a life and death thing, Dick!"

"We'll see," he replied.

* * * * *

At the appointed time, Dick and his son Tom, a vibrant eighteen-year-old, showed up at my house in Balmorhea. Tom was built like his dad—similar in size and musculature, but possessing the good looks of youth. He never fit in around McCamey, preferring the company of his dad to other kids his age. Because he worked all the time, Tom never participated in extracurricular activities like the rest of the kids.

"Need some new tires for your truck?" Dick asked.

"Yeah, I guess so," I replied. Tom unloaded two tires from Dick's pickup and rolled them to the back door of my house.

"Where did you get those?" I asked.

"The Co-op," Dick answered.

Later I found out that someone had driven a vehicle right through the front doors of the Bakersfield Valley Co-op and helped themselves to a number of items, among them several sets of tires.

We drove to Santa Elena and looked the area over, discussing how, when and where we were going to cross the marijuana. I told him I preferred coming out of the park during early morning hours. Typically, tourists leaving the park are headed out at that time of day and I liked to blend in. We decided to wait until night to cross the river. In the meantime, we drove back to Marathon, sat in a café and drank coffee. Later that evening, we returned to the campground, got a spot and waited.

After dark, the three of us walked to the river, struggling under the weight of the tires, and waded the river to meet my connection, who greeted us with a welcome smile. We inspected the marijuana and watched as he weighed each sack on a fish scale near the back door of his house. Unlike other towns in that region, Santa Elena had electricity, delivered through wires from the American side of the river. We left the tires and money in exchange for the marijuana. Tom and I each hefted a heavy tow sack full of marijuana to our shoulders, and the three of us walked back to the bank of the river. Dick stopped there and explained how he wanted to do the crossing.

"I am going ahead. You two trail me about fifty or a hundred yards behind.

If you hear shooting, drop the weed and run," he said. "I'm too old for running."

"That's crazy, Dick!" I exclaimed.

"That's the way it's going to be!" he said. Dick was in no mood to argue.

Dick crossed the river and then Tom and I followed. Dick walked about a hundred yards ahead of us, gun drawn and ready for action. We didn't bother sneaking through the dense brush but instead walked right out through the main path leading away from the crossing. I shook my head thinking, *This is all wrong*. Amazingly though, everything was quiet and we walked unmolested all the way to the campground. Once we got there, we hid the weed under the hood and chassis of Dick's pickup with duct tape and baling wire and waited for daybreak. We didn't sleep but sat and talked while Dick smoked one cigarette after another. The next morning, the three of us drove to my home in Balmorhea and unloaded the majority of the marijuana.

I never hid illegal merchandise in my house. Instead, I hid it in a nearby pasture covered with a tarp and camouflaged with branches and brush. I agreed to take the majority of the load to Plainview and sell it while they went on with a few pounds Tom hoped to sell in McCamey. I told Dick that I expected it would take about a week for me to sell the load.

We said our goodbyes and Dick and Tom drove off. And that was the last time I saw Dick Graham.

* * * * *

Plainview was flooded with marijuana when I arrived. One week turned into two before I could sell the load. Even then, I collected only half of the money and was forced to front the remainder to Jeff Aylesworth, who in turn fronted it to others. I spent the biggest part of this time locked in a motel, smoking joint after joint and waiting while a TV played in the background. This was the era of MTV—all music, all the time.

This was also the era before cell phones. Both Dick and I made it a practice not to call anyone when we were smuggling just in case our phones were being monitored. Rather than call Dick to let him know what was happening, I decided to make a trip with Jeff to tell him in person. I liked the idea of having both ears attached to my head. Jeff and I drove into Fort Stockton to the house of a friend and small-time dealer I knew to see if we could sell him a pound or two.

"Did you hear about that guy Dick Graham?" he asked me.

"No," I answered.

"Man, that guy went off. He killed a cop! We've had a manhunt going on around here like you wouldn't believe. All the roads blocked, helicopters flying around, ranchers and farmers driving around with deer rifles! You wouldn't believe how much shit he stirred up. They finally got him though. Down at Sheffield."

The only living eyewitness to recount the true story is Dick's son Tom, and I have never talked to him since the day they left me at my house in Balmorhea. My version is second, third or fourth hand. I'm sure it's wrong in some respects, but I will tell what I heard and let others divine the truth.

Apparently Dick was not content to get the money from our trip and reinvest it. He wanted to get a huge load for one last trip which would finance a move further south to Belize, allowing him to retire in a fortress he would build with the proceeds.

My father's Bakersfield Valley farm bordered the Pecos River and Dick knew the country well from having worked there. Dick had dug some large holes near the bank of the river in a secluded area of salt cedar trees and built plastic-lined wooden boxes in preparation for a load of stolen guns he intended to take from the Red Bluff hardware store in McCamey, which somehow had gotten on Dick's shit-list.

He and Tom went late one night, parked behind this store, scaled the building, cut a hole in the roof and entered the building. While Tom waited on the roof, Dick disarmed the burglar alarm and then prepared to relieve the store of a generous rack of weapons. What he didn't know was that there was a new silent alarm recently added to the store. Dick's presence set off the alarm.

Apparently the alarm was giving the owner and the police trouble, so they thought it was a false alarm. Four cops answered the call but only one entered the store. Dick shot the cop in the abdomen with a shotgun and later in the head with another firearm. The cop died. Dick escaped through the roof.

A gun battle ensued outside the store. Tom was grazed by a couple of bullets in the exchange. Dick and Tom leapt into their pickup, Tom driving and Dick shooting through the sliding rear window of the vehicle. Only one deputy was able to follow them from town and he had to follow at a distance to avoid getting hit by Dick's fire. When Dick and Tom got near my dad's farm, Dick told Tom to stop. Dick jumped out near a cattle guard, weapon in hand, while Tom continued on. The sheriff's deputy couldn't tell that Dick had gotten out and continued on after the pickup, only to run into a hail of gunfire when he reached Dick's position. He survived the incident but couldn't continue the pursuit when Tom came back and picked up his dad.

Dick and Tom traveled cross-country through right-of-ways for power lines until the pickup's tires went flat. They abandoned the vehicle and continued on foot, covering twenty or thirty miles.

Police issued an all-points bulletin. All roads leading out of the area were blocked. Helicopters circled above. The cops found his truck. Dick and Tom hid in some brush outside a gas plant near the town of Sheffield and waited for night to fall again.

They had no guns or ammo left and were armed only with knives. Dick told Tom to wait while he walked into the plant to steal a vehicle. He got the drop on the night watchman, who turned out to be a young man he knew—one of Tom's friends. Dick told the young man to give him the keys to his truck, but the kid instead tried to convince him to surrender. Dick wasn't the surrendering type. What he didn't know was that there was another night watchman present.

This second man removed a loaded twelve-gauge from his pickup and tried to capture Dick. But Dick was not going to be arrested. He moved toward the man with the knife. The man shot him in the chest. When Tom heard the shot, he came running. The man turned and shot Tom in the face with birdshot.

The shot that hit Dick left a hole in his chest about the size of an apple. Tom was farther away, but both eyes were hit and pellets also entered his lungs. The two were forced to lie on the ground. Dick died in his son's arms.

Tom survived but was left blind and was subsequently sentenced to thirty-five years at the Texas Department of Corrections for his involvement in the incident. His biggest crime was total and complete devotion to his dad, who was, without a doubt, a modern-day desperado. The highway patrolman Dick killed left behind a wife and five children.

A few weeks later I had the awful responsibility of going to Ojinaga to describe Dick's death to his widow. She had already heard, but my testimony brought finality to the rumors. I felt powerless to comfort her as she sobbed in my arms.

That was the last I ever saw of her.

MOUNTAIN MEN

I was a poor salesman. It takes a different personality to smuggle than it does to distribute marijuana. There were times when I did everything right in regard to getting the right product and getting it into the country but failed on the marketing end. For that reason, I spent a lot of time and effort trying to convince James Robertson to buy my marijuana.

Robertson and his associates were originally products of the Jamaican and Colombian marijuana days of south Florida and operated under the assumption that anything coming out of Mexico was garbage. They had an unlimited market and could move multi-ton loads. I found good marijuana in Mexico and took them samples, but they remained unconvinced. Finally, I talked James into going to Mexico so I could show him.

The plan called for him to fly to Mazatlán, bringing along his wife and kids. They were to check into the *Costa del Oro* motel and have a good time vacationing. My cousin Phil and I were to travel by land from my home in Texas in two vehicles. I brought my wife and children, too.

We met at the hotel and spent several days at the beach, going from hotel to hotel and from bank to bank, exchanging dollars into pesos. At the time, one dollar was worth about 250 pesos, so $20,000 turned into a large sack full of Mexican money. I arranged a meeting with my friend Daniel on my way through the mountains en route to Mazatlán. He didn't have enough marijuana to fill our needs and told me we would need to travel down the mountain on foot with a donkey into remote valleys with cash in hand. There we could buy from individual farmers—some with a few kilos, others with more—until we had pieced together enough for a load.

Phil drove James and me into the mountains and left us at midnight on the side of the road near the tiny village where Daniel lived. I told Phil to come back the following night at the same time. Daniel met us with a friend and two burros and we started down the mountain. Two rode—James and Daniel's associate—while Daniel and I walked, carrying flashlights. James and Daniel's friend rode between us, immersed in pitch-black darkness. From time to time, I was offered a chance to ride but refused. I had seen the trails we would be negotiating by day. They were steep, wet, slippery, and narrow—barely wide enough to accommodate man or beast. I was not about to entrust my life to a dumb animal on those slopes, not at night anyway.

Where Daniel lived in the mountains at the top of the trail, it was cool and sometimes even cold year-round due to the altitude. As we descended, though, it got warmer. At eight the following morning, we reached the bottom, surrounded by tropical forests complete with native bananas, avocados, and other tropical plants. It was like being in a steam bath. We hadn't seen a level piece of ground for eight straight hours of walking. The ride left James exhausted. I began to worry. We were going to have to walk back up that same mountain to get out of there. The journey had hardly fazed Daniel and his companion.

A group of men had gathered in anticipation of our arrival and were seated around a small fire I assumed they were cooking over—they certainly didn't need it to stay warm. All wore huaraches. All wore long pants of various colors but no blue jeans. Most wore hats and button-up shirts with a good many of the upper buttons unfastened revealing glistening brown skin, covering smooth muscled physiques common to those of Native American descent. Each man was armed. Most had semi-automatic handguns tucked into their pants with the shirttail pulled over to conceal them.

James and I lay down against a log near the fire and rested while these men talked among themselves. When a fox appeared on the opposite bank of the narrow valley, they all turned, drew their weapons and fired, sending a hail of lead in the poor beast's direction. Their skill with a firearm was considerable.

They killed him on a dead run at a hundred yards—with handguns. A man wouldn't have stood a prayer.

And there we sat, unarmed, with a sack full of money.

After a while, the men brought out their merchandise. James and I inspected each man's product—most had only ten kilos to sell. The marijuana was packed into poly feed or flour sacks. All was *sin semilla* with big nice-looking red-haired seedless buds, but the potency was mediocre compared to what the government-sanctioned fields in the north of Mexico produced. This might have been because of the seeds they used but was also due to the climate. The area where we found ourselves received plenty of rain—too much to produce potent marijuana. The product passed our inspection, nevertheless. If it hadn't, we might have been hard-pressed to leave with that sack of money. As it was, each man's merchandise was weighed and we paid cash for his individual contribution to the load. They received the money with grateful smiles and handshakes, happy to do business. What we paid for the weed was not much by our standards, but I am sure it represented a lot of money to these men—money they had no chance of earning by any legal means available to them. By noon we had put together enough marijuana to load the two donkeys. We then borrowed a mule for James to ride back up the hill. It was evident both to Daniel and me that he would never make it out on foot.

Up the trail we walked and walked and walked. About five p.m., James began to complain.

"Ask him how much farther it is."

"Poquito más," Daniel replied.

"Good. I don't think I can stand much more of this."

I was struggling terribly to keep going too, but didn't say anything. I tried to look only at the next step, taking them one at a time. When I did venture a look up the mountain, the distance we had left to go appeared staggering. Another hour passed. James demanded we stop. He was getting to the point that he couldn't hang on to the mule any longer. Thinking it might be easier to walk, he dismounted. He lasted all of twenty minutes before falling to the ground and refusing to move.

"Just a little farther," Daniel said.

"That's what he said last time."

"I'm just repeating what he said," I told him.

James and I fired up a joint, sharing it with Daniel and his friend. After smoking, James remounted the mule and our trek began again. We walked and rode through dusk into darkness again. James struggled mightily to stay on the mule's back. I followed closely behind, occasionally grabbing the mule's tail to help get up a particularly steep spot. Daniel blazed the trail ahead, and his buddy brought up the rear.

About nine, we stopped again.

"How much farther?" James asked again, a note of desperation creeping into his voice.

"Just a little farther."

"That lying son-of-a-bitch has been saying that for three goddamned hours. How much more is just a little further?"

"Just over that hill," Daniel replied. *"Poquito más lejos."*

We smoked another joint and ate some *chicharrones* and *carnitas* along with a few corn tortillas Daniel's wife had sent. After the food and yet another joint, James was ready to try again. I was really exhausted too, but I told myself as long as that Mexican kept walking, I would keep going. And I did. Just about the time I thought I couldn't go any farther, we reached the top. It was almost midnight, twenty-four hours after our walk began. We had been walking for eighteen of those last twenty-four hours without any sleep. And without crossing a hundred yards of level ground the entire way.

Phil pulled his maroon van to the side of the road. We descended from the forest-covered side, slipping and sliding down the wet slope with sacks in hand. After everything was loaded, we drove off. About ten miles down the highway toward Mazatlán, I knew of a remote logging road. We pulled in there and repackaged and concealed the marijuana in hidden stash compartments in the van.

Around six the next morning, we pulled into Mazatlán, weary and disheveled. After a day's rest, James and his family flew back to Oregon. I drove north in the loaded vehicle. Phil, Cheryl and my children followed behind in another. We rendezvoused in Durango and spent the remainder of the night in a motel there. I saw a street vendor with a bicycle-powered cart and a crowd of people around waiting for something to eat. Thinking whatever he had must be good, I bought a sack full of tacos and took them back to the room. But they contained some unidentifiable body part that looked like a bunch of big jugular veins with the surrounding fat still intact. Needless to say, we went to bed hungry that night.

The next day we continued north and then parted company.

Cheryl and the kids drove to Piedras Negras and crossed legally into Eagle Pass while Phil and I went through the desert and crossed at La Pantera. Then we drove nonstop all the way to Medford, Oregon.

I never could get James to do another deal in Mexico. He told me, "You get the dope up here and I'll buy it."

I never walked to the bottom of that mountain again. Once was enough.

* * * * *

James was not impressed with the quality of the marijuana we got from Daniel, so Phil and I decided to take some of the good lime-green *sin semilla* Oscar had to Oregon. We also got some for Kenny Schulbach. We loaded the marijuana south of the Big Bend and drove to Medford again. Phil and I hired an additional driver for these trips, a Vietnamese fellow we called Mr. Li. Jeff and Kenny drove along behind with another fellow, also switching drivers. While one person drove, another rode shotgun and another slept in the back of the van. Kenny took his share to a connection he had in Northern California. As soon as Phil and I arrived in Medford, we unloaded and headed back for more.

When I first met James, he refused to use or sell hard drugs. Time had changed him. He no longer looked like a hippie but instead like a white-collar businessman. And he used and sold cocaine. On one of these trips, James had some. The pile I saw contained thirty kilos worth $56,000 a kilo wholesale—fresh off of a boat from Colombia. I'd avoided using much of this stuff at this point in my life, but when he gave me five hundred grams to take home, I immersed myself in it. I spent the next six months high on cocaine, six months I was fortunate to survive. I quit cocaine to save my life—without having to be busted. It was either quit or die.

Way too much cocaine is never enough.

THE BEGINNING OF THE END

I first bumped into David Regela when I was driving north out of Big Bend National Park with a small load of marijuana and a carload of people. As usual, we had gone to the park posing as tourists, purchasing approximately thirty pounds of marijuana to sell and show others in order to gather money for a larger load. The passengers that day included Crocket and Cubano from Balmorhea and Jeff Aylesworth from Plainview.

The car was Jeff's two-door Impala. We had removed the back seat and the interior panels. This opened up a hole above the rear fender wells on each side of the body of the car which we then loaded with marijuana. We reinstalled the panels along with a dab of super glue on each of the retaining screws and then we replaced the back seat.

We were driving along when a pickup passed us and blocked the road in front of us while another, flashing lights, stopped behind. Two men, one from each vehicle, got out, weapons drawn, and approached our car, ordering us to keep our hands in plain sight. The man from the vehicle behind us identified himself as David Regela.

The first thing I noticed about the man—after the big gun in his hand, of course—was his hare lip. This condition made him lisp when he spoke. The second thing I noticed was that he was a confident, aggressive type—what some might call a cocky son-of-a-bitch. We were ordered out of the car and forced to wait in a group behind the vehicle. Both Regela and his partner took turns searching our vehicle. I watched, knowing that if that back seat didn't come out, they wouldn't find the marijuana. Regela did, however, find some joints I had hidden under the seat for the ride home.

The night was cool under a West Texas sky full of stars. They searched and searched but they missed the load. I have a vivid memory of Regela standing with a fistful of pre-rolled joints held up in front of my face, lisping, "If this is all you have, you can go. We're not interested in little amounts."

Of course, I lied and admitted that it was. Crocket had the audacity to ask him for at least one joint for the road. Crocket was in his normal state of mind—drunk—and fairly insistent about it. I felt like strangling him. Regela laughed it off and let us go anyway. I wondered at the time if he kept the joints for his own supply.

* * * * *

The second time I met Officer Regela was several years later, in 1985, shortly before my first American arrest.

Phil, Kenny Schulbach and Mr. Li drove from Fort Worth and met me in Fort Stockton. We rented a room there for Mr. Li, who didn't want to go near the border. The rest of us proceeded to La Pantera in two vans. We drove through the river and set up a camp several hundred yards beyond the crossing. Oscar Cabello was scheduled to meet us there with a large load of marijuana.

Kenny was a gun freak. We had with us, among other things, a CAR-15 fully automatic rifle. A CAR-15 is basically the same weapon as a military issue M-16, only it has a telescoping stock and is more compact than the standard army issue. Without advising Phil or me, Kenny got out this rifle and fired it in the full automatic position, spraying bullets indiscriminately into the air. Later, I found out that the gunshots startled some tourists on the American side of the river, who then drove off and reported the incident. Oscar didn't show at the designated time so we moved camp farther into Mexico, about ten miles or so, to an old line shack I knew about that would be on the path from Piedritas to La Pantera. We killed time there, doing additional target practice and waiting. The next day Oscar showed up with only eighty pounds, telling us that the shipment we expected had not yet arrived.

I offered to wait in Mexico for the remainder of the load while Phil and Kenny took one of the vans back to Fort Stockton to tell Mr. Li what was happening. A couple of days passed. I decided to move to another location. I didn't like to stay in one place too long. Oscar still hadn't received any more merchandise and I had no idea what had become of Phil and Kenny. My new location was near an abandoned fluorite mine farther south, toward the town of Piedritas.

I heated water for coffee and cups-a-soups in the van, using a small propane burner that also served to heat the interior of the van itself when the cold night air arrived. And waited.

Around mid-morning the next day, Phil came driving up in his maroon van.

"You're not going to believe what happened," he began. "Do you remember a Customs Agent by the name of David Regela?"

"I don't think so."

"He has a harelip . . ."

"Oh yeah, I know that guy. He stopped me a couple of years ago."

"Well . . . he caught us."

"What do you mean, he caught you? You didn't have anything on you."

"You don't have to. They have these laws called conspiracy laws now. But it all worked out. He let us go. I worked out a deal to pay him in exchange for his protection."

"Wait one fucking minute here! You made a deal with an American cop to bring marijuana into the country?"

"Yes. Take it easy, Don."

Phil continued. "This may be just what we need. He can tell us when the way is clear. We can bring larger loads because we won't even have to hide the stuff."

"I don't want anything to do with this," I told Phil.

"Hear me out. You really have no choice. I already paid him for one load. If you don't want to do it, then I will. He already has all he needs to bust us for conspiracy to import, but he isn't going to. He knew all about you. He's been watching you for over two years. What he didn't know before, he knows now. Besides, I already made the deal. We have to do this."

"What do you mean we have to? I don't have to do a goddamned thing!"

Phil had described every aspect of our operation to the Senior Customs Patrol officer of the entire Big Bend region.

Here's what Phil told me:

Phil and Kenny had returned to Fort Stockton without being stopped. They rented the room for more nights and explained to Mr. Li what was happening.

They also picked up supplies for our wait and started back toward the river. Somewhere on the river road leading toward Talley Crossing, a wheel bearing went out on the van. Kenny hitched a ride with a passerby to get parts or a wrecker to fix the van. Phil waited behind with the van. A park ranger came along, the same one who had once called a wrecker for us when we crashed coming out of the park, and questioned Phil. I guess he recognized either Phil or his name and called it in. Regela responded to the call. Soon afterward, the ranger and Regela returned and searched the van, finding a half ounce of marijuana, two handguns and several thousand dollars in cash either in the van or on Phil's person. During the search, Kenny arrived with a wrecker.

Regela arrested both Phil and Kenny on the spot for conspiracy to import marijuana. The wrecker driver was sent to the park headquarters with the van while Phil and Kenny were cuffed behind their backs and loaded into Regela's car. Regela, the park ranger, Phil and Kenny then drove to La Pantera to wait for the load Regela assumed would be coming. At the crossing, Regela placed Phil and Kenny on the bank of the river, sitting, with hands cuffed. He waited behind with a weapon, using them as a shield. Kenny complained bitterly about this—to no avail. Hours went by and no one showed. As it got darker, Regela decided that he was going to have to take them in. He drove. The park ranger rode in the front passenger side seat while Phil and Kenny rode in the back seat. Somewhere before reaching the place where the park ranger had left his pickup, Regela stopped the car, opened the trunk, removed an automatic rifle and chambered a shell. He then opened a door and ordered Phil out, leaving Kenny and the park ranger sitting in the car.

Regela led Phil to a spot in the desert where there was a depression in the ground. Phil said it resembled a gravesite. Regela shoved him to the ground and then threatened him at gunpoint, saying, "We can just leave you here for the buzzards to eat." A conversation ensued. Regela described himself as a man with marital troubles who had either one or several ex-wives and a current girlfriend with expensive habits. He told Phil that he had once been an air marshal. He told him he made only forty grand a year at his job and that that wasn't enough to support all his past and current relationships and that he needed more money. He told Phil he had been looking for someone he could work a deal with, providing protection and information in exchange for money. He said he had everything he needed to bust us under conspiracy laws, but suggested that it might be possible to work out a little sweetheart deal if Phil was interested and was qualified for the job. He told Phil that for this thing to work, Phil would need connections large enough to satisfy his demands, since Regela would like to earn in the neighborhood of thirty grand a month. If we couldn't operate on that scale, it wouldn't be worth his time.

Phil assured Regela we could. They talked more, sharing stories and information. Regela then led Phil back to his car. After dropping the park ranger at his vehicle, Regela resumed the conversation in Kenny's presence. While Phil and Regela firmed out the deal, Kenny listened, saying occasionally that he wanted nothing to do with it. At the park headquarters, Kenny fixed the wheel bearing while Phil and Regela continued their conversation and shared a joint or two in the process. Phil told Regela that I was waiting on the other side of the river with a load. He gave Regela a down payment, got into the van—which by this time had been repaired—and started toward Mexico. After crossing into Mexico, Phil had been unable to find me and had gone on to Piedritas to find Oscar. Kenny remained behind at the park headquarters.

After hearing Phil's account at my campsite near the fluorite mine, I replied, "I don't like this."

"What choice do we have? I told him everything—how much we do, when we do it, how we do it—everything. He has all he needs to bust us."

"What are you talking about? You didn't have anything on you! How can he arrest you and charge you for smuggling without any dope?"

"You don't have to get caught with dope to be charged with conspiracy," Phil advised me. He continued, "Here's how Regela explained it. If four guys sit around and talk about burning down a building and make plans to do so and then three of those guys go home and never do a thing but one guy goes and burns the building or even makes a move toward doing so, then all four are guilty of conspiring to burn down that building."

"I don't know about that," I replied, suspiciously.

"I have already paid the guy!"

"What do you think, Oscar?" I asked my friend.

"I know of the guy, *El Tartamudo* (the Harelip) we call him. This might be real. If it is, it would be a good thing, but who knows."

I told Phil, "If all he really wants is money then *you* can go give it to him. I'm *not* going to meet with him. I *am* going to smuggle this load and head straight to Fort Worth. If you want to give him money after the fact and let him know what happened, that'll be up to you."

"He's not going to like this! We're supposed to meet with him."

"Screw him."

I drove through the river and sped through the park roads. Once I hit pavement, I drove a hundred miles an hour all the way to Fort Worth. If anybody was following me, I was going to know it. Phil came along behind, drove back to the park headquarters and delivered more money to Regela after the fact. Then Phil drove to Fort Stockton, picked up Mr. Li and Kenny—who had been released and found a ride to the motel—and the three returned to Fort Worth.

Phil told me Regela was adamant that I meet with him. I said no. Over the next couple of weeks, Phil and Regela talked over the phone. Phil went so far as to suggest that the safety of my family might be compromised if I didn't cooperate. Regela told Phil that in order for this thing to work, a great deal of mutual trust would be required. Eventually, I was persuaded.

We met in Phil's van near the park headquarters about a month after Phil and Regela's initial meeting. We didn't know it, but Regela wore a wire.

Regela's story didn't sound much like what Phil had described. I listened, wondering about the differences but saying nothing. Regela described the arrangement Phil and he had made and asked me if I wanted in on the deal. I agreed to his terms. Because there was no marijuana left near the border, I told Regela that we would have to go farther into the interior of Mexico to acquire a load. We agreed to give him one day's advance notice of our return. We also gave him money.

We bought and smuggled a load out of the mountains of Durango and Sinaloa and smuggled it into the park, crossing once again at La Pantera. I decided, once again, to go on without meeting Regela, at least until after the fact. I didn't give Phil an opportunity to call Regela before our return. Phil was unhappy, to say the least. When we got to the store near the park headquarters, I went in to buy some snacks. Phil got out to call Regela without telling me. Moments later Regela drove up. When he drove up, the plan to continue on without notifying him went out the window. He was mad that we hadn't called in advance.

Regela told us he wanted to inspect the load to make sure we weren't sneaking more weed into the country than we had agreed to pay him for. The marijuana was concealed in a false bottom in the bed of the van. We drove back to Dugout Wells and met him there. On the way, Regela called in on his radio and arranged a roadblock complete with a mob of heavily armed law enforcement officials.

We drove out of the road to Dugout Wells, turned north on the road leading to Marathon and obediently drove into the trap waiting for us like sheep headed to slaughter. We were then arrested in a ridiculous display of overkill. Looking at the firepower amassed for our arrival, you would have thought they were arresting an entire invading foreign army.

Afterward Regela apologized to us, saying, "If there was one crooked bone in my body, I would have liked to work with people like you, but there isn't."

After I was busted, I read transcripts of the conversation in the van when I gave him money. It was apparent to me that Regela, the only one at the scene aware that the conversation was being recorded, was painting things favorably for his position. It was also evident that he had help. Several times during our conversation, things I said that might have looked good in our defense somehow were garbled and unintelligible on the tape or totally missing.

One instance comes to mind—when Regela asked if we could get any cocaine. I told him I no longer used or sold cocaine. I told him I had known people whose lives were really messed up by cocaine and wanted no part of contributing to their downfall. Phil also stated similar views. None of this appeared on the tape.

I wasn't the only one with doubts about Regela's sincerity or honesty. Shortly before my release from prison, I was taken out and interviewed by two internal affairs officers from the Department of the Treasury. The nature of their questions led me to believe that Regela had done something else to fall out of favor with the Customs Agency. I never found out exactly what that was, but I did hear rumors. The rumors were bad.

The internal affairs officers did say Regela no longer worked for Customs and was under investigation for crimes. Someone else will have to tell that story. I would love to hear it. It's possible, perhaps even probable, that he always intended to bust us and never really considered taking bribe money. All the money was accounted for. Maybe he just improvised after arresting Phil and Kenny with no evidence. To be sure, cops have the disadvantage of following strict rules smugglers are unencumbered by. Years later, while reading Terrence Poppa's *Drug Lord*, I read accounts of Regela's meetings with Pablo Acosta that added to my suspicions, but, once again, there exists no absolute proof of wrongdoing in these accounts.

I do know that Regela is now, or at least was in the not-too-distant past, still alive, and I have heard from a reliable source that his days as a Customs agent left him an emotional wreck.

In defense of drug cops, I will say this. The task of those assigned to stop the flow of drugs into this country is an impossible one. They win the battles; they lose the war.

JAIL

After our arrest, I was loaded into a car with a DEA agent by the name of Eino Hella. Phil and James Risenhoover, an unfortunate driver who just happened to get caught with us in his first attempt to smuggle marijuana, rode with Regela. I rode handcuffed alongside agent Hella. A Mexican national who had been caught earlier in the day, also with marijuana, rode in the back seat. On the way to Alpine, we talked.

I asked, "How much time do you think I'll get?"

"Probably ten years," Eino said.

I replied, "That's ridiculous. How in the hell can you justify putting me in jail for that length of time—for marijuana?"

He went on to explain to me why he thought marijuana should be illegal, arguments I had heard before. "It causes cancer. It stays in your system for long periods of time. It lowers your IQ. Users become lazy and neglectful. It causes short-term memory loss. The rest of us pay for loss of productivity and absorb the cost of medical treatments related to its use with higher insurance premiums."

"What about the combination of alcohol and tobacco? They do everything you say marijuana does—if not more—to harm our society," I said.

"If I had anything to do with it, they would be illegal also," he answered.

This was the first satisfactory reply I had heard to my argument. I can still feel the sting of his words—like a slap to my face or the crack of a board on my ass.

While I disagreed, I understood. I still disagree with the inequity of our laws, but I still understand.

* * * * *

The door to the cell is opened. James, Phil and I are ushered in and the heavy metal door slams shut behind. The room is full of wetbacks. The cell has no windows. It measures maybe ten by twelve, with fifteen of us inside. The floors are cement and it's cold, artificially so. It's hot outside. Scared, innocent eyes stare up at us. Men sit around the edge of the room and in the middle. We stand. Finally, they squeeze against each other to make room for us to sit.

"Gracias," *I say.*

One of them asks if I speak Spanish.

"Yes," I tell him. "What did you do?" I ask. I know the answer, but I ask anyway.

"We were caught by the Border Patrol," he tells me.

In other words he didn't do a goddamned thing. I tell him I'm sorry.

He looks confused.

"What will they do with us?" he asks.

"Don't worry. They'll turn you loose," I tell him. But I don't know.

"What did you do?" he asks me.

"We got caught with marijuana."

His eyes open wide. Now it's him who's sorry for us. "What will they do with you?"

"They'll lock us up for a very long time, I think," I tell him. I don't yet understand those words. It will take years to fully understand them.

Hours later, guards strip-search us and take our clothes. We're given white jumpsuits and plastic slippers that slap the concrete floors when we walk. They also give us a sheet, a wool blanket and a plastic-lined mattress, about two inches thick. We're led to another iron door. It opens to reveal twenty-three sets of eyes. The cell is designed for twelve. We're squeezed in and the door is slammed behind us. We stand, wondering where to put our stuff. There aren't any open bunks. Finally, one of the men, a Mexican from Hobbs, says, "You'll have to sleep under the table." We walk to the end of the cell and unroll our mattresses. Our feet lay under a table lined with men playing cards. Day and night, the game goes on above us.

Welcome to the Reeves County Jail, Pecos, Texas. You won't easily forget this place.

They tell us that in three days we'll be going to court. We haven't bathed. There's only hot water in the shower for a limited time and there are twenty-six of us. One Bic disposable razor is provided for twenty-six faces and then picked up an hour later. At five a.m., breakfast arrives—a glass of warm powdered milk with rice in the bottom and coffee. The line forms. We don't eat much. That will change. Soon we will look forward to that rice.

When we go to court, we're wearing the same dirty smelly clothes we were caught in and taken across the street, chained together. Once again, we're stuffed in a room with wetbacks. I recognize a few faces. I'm dismayed when I find out that some of the men are to be held—as evidence. They'll sit in jail until the one deemed to be the coyote goes to court. This means at least sixty days.

Nothing happens. We go back to the jail and resume our life in white jumpers. I talk to the man from Hobbs. He picked up two men—relatives—coming from Mexico. He was stopped and it was discovered his passengers had entered the country illegally. His pickup is confiscated. He is charged as a coyote. He has done sixty days in jail and lost his job. His wife is behind on her rent and soon may be kicked from her home along with his kids, all of whom are American citizens. He will lose his green card and be deported—after he does his time. His friends are also held—as evidence of his crime. They're deported but he remains locked in jail for this awful deed, this awful crime of helping a relative looking for honest work.

I look around the cell. Out of twenty-six of us there are maybe two or three real criminals—violent predators. All the rest of us are in for drugs or smuggling wetbacks.

The door opens and three more men dressed in white jumpers and plastic sandals come in with rolled-up plastic mattresses. There is one place left on the floor, right between the two tables. The heavy iron door slams shut.

★ ★ ★ ★ ★

When the indictments came down, I was facing a variety of charges, including conspiracy to import marijuana, possession with intent to distribute marijuana, conspiracy to bribe a federal official and bribery of a federal official. I had a court-appointed lawyer, a guy by the name of Rogers. Phil had a lawyer by the name of Barclay. Both worked out of the same office. We told them our story. I told Rogers I would plead guilty to the importation and the possession charges but not to the bribery. The importation carried a maximum of five years and the possession fifteen.

"No way! This is a clear-cut case of extortion! Don't even think about pleading guilty."

I said, "But I *was* smuggling marijuana before David Regela came along. I am guilty of that. The bribery is another matter."

He persuaded me that we needed to fight the case because of Regela's "illegalities." He also convinced me that I needed to pay him in order to get a proper defense. Now *I* was the one wanting to have *MY* day in court. A series of motions were filed on our behalf: a motion for discovery and a motion to disallow evidence. The government presented its evidence. Hope began to grow in me. Clearly, Regela had overstepped his bounds.

The day arrived to pick the jury. But—after milking us for thirty grand our families could have used—Barclay and Rogers arrived with plea agreements in hand. My dad was near broke and had spent his last money helping us pay them. Phil's mother gave Barclay a family heirloom she had kept for years, a set of candlesticks that had once occupied a spot near the throne of an English king.

The agreement consisted of two charges—conspiracy to import and conspiracy to bribe a federal official.

"Hell no!" I told him. "I'll plead to the possession with the intent to distribute charge and the importation, but not to bribing that bastard."

"The possession charge carries fifteen years. These two charges carry five apiece and I can virtually guarantee you they will run concurrently. You will go to a minimum-security prison and be out in twenty months. Take it."

"No! At least I committed those other crimes. I don't want to plead to the bribery. I would never have bribed him. I was conned into doing that."

Rogers said, "I won't be a party to your hanging. Either you take this or I am going to recuse myself from this case."

I took the plea bargain and regret it to this day. Before accepting the plea, the judge asked if we had been threatened or coerced. I told him that I had not been—not directly—but that implied threats were delivered to me through my cousin. When Phil was asked about this, his reply was so weak that it was almost inaudible. He just wanted to get the whole thing over with.

I was set up by a whiskey-drinking cop. Through my cousin, he influenced me to commit a crime I was not predisposed to commit, at least here in the United States, where the good guys are supposed to be good and the bad guys, bad. The entire legal process that sent me to prison was a farce. Our fate was most probably determined in a bar or a restaurant, between the judge, the prosecutor, and our defense attorneys, who all happened to be good friends.

David Regela was good at putting smugglers in jail. He says he wasn't crooked. The nature of his job required being a proficient liar. To this day, I don't know what to believe. He smoked marijuana with my cousin Phil. Kenny Schulbach witnessed this. Phil says Regela drank heavily—Wild Turkey whiskey.

Cops, who drink and smoke on one hand and imprison others for the same, merit no respect in my book.*

* Years later when I read Terrence Poppa's Druglord, I saw that Regela went down to Mexico and partied with Pablo Acosta—one of the most violent, murderous smugglers of the era. But he busted us.

ESCAPE FROM PRISON

On my first day at Big Springs prison camp, I was searched and issued clothing and then led to our "cell." To call what we lived in at Big Springs a cell is a lie. The facility was as nice as most college dorms. I was sent to the Sunset Unit and then to the guard on duty for a brief orientation. A short muscular black guard greeted me and several other new inmates. He looked vaguely familiar. He did a double-take at me.

I read his nametag. *Walton. Where do I remember that name?* I'm 6'1''. Walton stood about 5'6''. As I brushed by him, he seemed intimidated.

From that day forward, Walton decided to make my life miserable. In prison, a guard's biggest job is to count. Every time you turn around, they are counting you. Walton had a hard time with numbers—I guess there were just too many of them for him and he tended to get them confused. Any time the numbers didn't add up, a recount was in order. When Walton was on duty, a recount was a certainty. Inmates rightfully began to connect Walton being on duty with the count coming out wrong and would shout, "Recount" when they saw him. Walton figured this was my fault.

The official daily count was conducted at four p.m. every day. It was sent into the Bureau of Prisons headquarters. Each of us was required to stand in our doorway for this count. We were supposed to remain standing in the doorway until the count-cleared signal was sounded. One day Walton was counting—a bad sign. We would probably be left standing for an extended period until he got the numbers right. One of my roommates, an inmate by the name of Long—in prison for bank fraud—decided to take a chance and bathe after the guards had gone by. I warned him about doing this with Walton on duty, but he took his chances anyway. About the time Long got good and lathered up, the door at the end of the hall burst open and Walton yelled, "Recount!"

Long came running up to the door with shampoo in his hair, hastily wrapping a towel around his body. Walton came to the door and lost count. He looked at Long—then at me—he'd have been white-hot if his black skin hadn't covered it up. Nothing could hide the rage in his eyes.

"Why did you let him do that?" he asked me.

I looked at Long, then at our other roommate, a slender tattooed Mexican heroin dealer from Robstown, Texas, who shrugged his shoulders.

"Why don't you ask him?" I answered.

You'd have thought I called his mother a whore. If I thought Walton was on my case before that, I was wrong. I began to get angry at him too and came very near to hitting him once when he jumped me because another inmate's radio was too loud—Long's, once again. I searched my mind trying to figure out why this guy hated me so much and then one day it came to me.

* * * * *

Football in West Texas is almost a religion, and in the seventh grade I decided to be part of that religion. At the time, "church" was held on the practice field at San Jacinto Junior High (home of the future President of the United States—Dubya, they call him). I stood 5'6'' tall and weighed all of 120 pounds, but I knew how to use what little body I had. I played mostly defense and got plenty of tackles. I remember sacking the quarterback of our offensive team on three consecutive plays in practice one day.

The stars of our team, however, were a pair of young black men, one called Caveman and another who went by the name of Zachary. Both stood about 5'6'' tall and weighed about 180. Both could run—much faster than me or anyone else on the team. Both were far stronger than the rest of us and both played in the backfield. Every seventh-grader's nightmare was the day you had to tackle one of these guys. One day I guess the coach got tired of seeing me, this skinny kid, getting through the offensive line and tackling everybody and decided to pair me against these two fellows in a drill—two on one. I charged fearlessly toward them, got knocked backward and landed on my back about five yards downfield. I got up with my ears ringing and fighting for a breath, but I did get up and I didn't forget.

Later the coach lined two players opposite each other with about a ten-yard gap in between. He'd throw the ball to one, who would then charge the other whose job it was to tackle the ball carrier. The coach lined me up against Caveman and Zachary, one at a time. Most of the other players would hit them on the thighs or the body and end up flat on their backs as they ran by with the ball, cleating them in the process. I developed a method now referred to as spearing—I would plant my helmet right into their shinbone with all the force my skinny little body could muster. I'd get up, ears ringing and dizzy, but they'd be on the ground too. And it hurt. I know it did.

One day, our sadistic-minded coach dreamed up another drill. A receiver, either a back or an end, would run out for a pass toward a waiting defender. The coach would throw the pass and let them fight to see who could catch it. I drew Caveman. The pass was thrown. Normally I would have speared the guy, but on this particular day I decided to try another tactic. While he had all his attention focused on catching the ball, I caught him right in the neck with an extended arm. I guess by modern rules this procedure, known as clotheslining, is illegal. On a West Texas practice field it wasn't and it worked. Caveman went down and almost out. Next time up, I drew Caveman again and, once again, I clotheslined the guy. This time he didn't get up right away and when he did, he had tears in his eyes. And he was mad. He tried to complain to the coach and got nowhere with it.

* * * * *

Then it hit me: *Caveman Walton.* After that, whenever Walton came around, I made sure everyone knew his real name. "Caveman," they'd holler, usually followed by "Recount!" This didn't help get him off of my case—on the contrary, the abuse intensified, but it made me feel better.

I remembered Caveman Walton as a big guy. That's why I didn't make the connection. He hadn't grown an inch or a pound since the seventh grade. For all I know, he might have been eighteen at the time. Being a star, I think he figured he'd go pro and didn't need to worry about minor things like learning how to add. When things didn't work out with the career in football—and, more than likely, with a number of other jobs requiring competence—the Bureau of Prisons came along and rescued the man. It doesn't take a rocket scientist to be a prison guard, although they do prefer it when you can count.

* * * * *

One day I was called in to take a psychological exam along with a test designed to check our level of education. After the test we were forced to interview with a psychologist. He asked me lots of questions and then dismissed me.

"Wait a minute," I responded. "I have a question for you."

"What is it, young man?"

"Now that you've tested me and asked all these questions, what do you think? What can you tell me about myself?"

"Well, for one, you are a very angry young man."

"You got that right!" I said. *Motherfucker!* I thought. *What do I have to be pissed off about? Seven fucking years for smuggling marijuana. No, let's rephrase that—four years for smuggling marijuana and three more for bribing a federal official—a federal official I didn't want to bribe, who forced the crime upon me, a crime I was not predisposed to commit—a federal official who was a lying, extorting, piece of dog shit.*

Goddamn judge looking at me through bloodshot eyes from the booze he drank the night before. Goddamn marshals smoking cigarettes non-stop. Redneck guards too sorry and lazy to get a real job—also smoking cigarettes and nursing hangovers. My wife and kids out on the street, God only knows where.

I think it was fair to say he had me pegged.

Having turned myself in to this goddamned place voluntarily was the worst part. There was next to no security at the minimum-security facility. They had us jailing ourselves! I walked around looking at the glaring lack of security and asked, *What is wrong with us—staying here like a bunch of obedient dogs?*

* * * * *

We were forced to take a job in the camp. So I went to work making mailbags for the U.S. Postal Service. They even paid us for this privilege. The starting wage was twenty-two cents an hour. If you stuck with it for a couple of years without getting transferred to another prison, you could get paid up to a dollar an hour. Of course these huge wages created loyal, hard-working, devoted employees.

I rewarded my Unicor bosses by sewing like a maniac, making as many bags as I could in an eight-hour shift. After a couple of months, I made a thousand bags a day and had been given a raise all the way to forty-four cents an hour. I would even have leftover money to send home! Some of my fellow inmates were not happy about me working so hard because they produced all of three hundred per day apiece and felt like I made them look bad. If these other inmates had been able to see what was going on in my head, they wouldn't have been so inclined to think I was a brown-nosing company man. I was pissed off at the world and immersed myself into my work in a state of mind just short of controlled rage, working until exhausted and then heading to the gym to fight the iron pile until time to sleep. Someone said something derogatory one day and an older, newly arrived inmate defended me, saying, "Let him be. He's just trying to do his time."

Lee Cross understood I was just doing time the best way I knew how. I couldn't have cared less about making my bosses happy. Lee was about sixty to my twenty-nine. He told me he had a fifteen-year sentence on a cocaine charge that reeked of entrapment. All he had done was make a call connecting a buyer, who turned out to be a FBI agent, with a seller he knew. The total amount of cocaine was one kilo—a kilo Lee never even saw nor stood to make any money off of.

Lee appeared to be a kind and intelligent man. I was appalled that he got such a heavy sentence for such a minor offense—in essence, a death penalty, for a man of his age. I immediately felt sorry for him, but Lee wanted none of that. The man was tough as boot leather, both physically and mentally. Lee worked at the Unicor factory for one reason only—he smoked cigarettes and liked to drink coffee. He had no one to send money to and needed the work to support his habits. It also helped him do his time.

Inactivity is your worst enemy while locked up. The easiest way to survive prison is to stay busy doing something—anything other than thinking about the world outside the wire. Those who don't learn this lesson, who stay on the phone, constantly worrying about problems they have no control over, do hard time.

Lee smiled a lot—his big blue eyes were magnified through the thick lenses of his glasses. In spite of the fact that he smoked, he had a good physique for a man his age. He bathed routinely and groomed himself each day as though he were going out to a real-world job where such things matter. He didn't lift weights but walked—each and every day he walked and walked and walked—thinking. If he wasn't walking, he was reading, and not junk. He liked authors like Kurt Vonnegut and Ernest Hemingway.

I don't remember exactly how the idea to escape came about. I like to think it was my idea, kind of like a guy likes to think it was his idea to ask a girl on a date when the girl really did the asking without having said a word. The truth was, Lee needed to escape and I was pissed off at the world, especially the part of it that had me locked up at the time and I wanted to resist. Escaping seemed a logical way to do just that.

* * * * *

I hadn't smoked any marijuana for the three months I'd been at the prison camp. I finally broke down and smoked a joint. The next day, I got called in for a urine test. Maybe this was a coincidence, but at the time I didn't believe so. A dirty urinalysis would mean more time before I would get paroled. It would also mean I would be transferred to La Tuna, a real prison, where people get shot trying to escape. If I was going to leave, it had to be done before the results to that test came back. To be honest, the effect of the marijuana probably had something to do with my change of heart. Before that, I was an angry young man—but I was also determined to do my time. The minute I got high, my attitude changed.

* * * * *

Lee invited me over to the Sunrise dorm to meet a couple of inmates interested in becoming marijuana smugglers. The first guy was David McCasland. David was serving the last week of his sentence for stealing semi-trucks loaded with hanging meat. The second guy was a marijuana smuggler by the name of Jimmy Hobbs. He was looking for a place in Mexico near the border where he could land and refuel with loads originating farther south, or if not, buying from my source near the border.

David owned a farm in Tucumcari, New Mexico, and had been saddled with the awful task of making money growing crops—a cause I readily identified with. He needed to make a lot of money quick or risk losing the farm. When he was in prison, he had met the likes of Hobbs and Lee and discovered the tremendous amounts of money available to those willing to smuggle dope. He was a licensed pilot and his farm near Tucumcari would make for a good place to land planes. Lee was also a pilot, although his license had long since lapsed. No matter—they don't ask for licenses to fly planes loaded with marijuana.

The plan took shape in McCasland's room. McCasland would get out, check in with his parole officer and then return the next night to pick us up. Hobbs got out several days before McCasland and would work on getting his connections down south in order. My job was to go with Lee to Piedritas. I didn't want to smuggle any more dope into the country, especially since I would be a fugitive. Instead, I would connect these three with Oscar for a tiny cut of their action. A one percent commission on the quantities they were talking about would be a lot of money.

We would leave on a Sunday at 7:00 p.m. That left us three hours to get out of the country before the ten o'clock count. Rather than go toward Mexico immediately—which I figured they would expect if they looked at my record—we decided to go back to Tucumcari with David to hide out at his ranch for a time and then head south later.

As the hour neared, Lee and I stood watching the road near the prison from the balcony of the Sunset Unit. As usual, the patrol car driving around the perimeter made his pass. Another car drove by with the interior light on precisely at 7:00 p.m.—our sign to head for the fence. While he drove to a predetermined point to turn the car around, we walked quickly down the steps, across the yard and by the chow hall and then took off running for the fence through the dark night. We scaled the fence and ran down an embankment, arriving at the exact moment McCasland's car did. We opened the door, got in and drove off.

A stroke of good fortune came with all of this. Caveman Walton was on duty when I escaped and I got to mess up his count—one last time.

As we drove through the night toward Tucumcari, my watch read 10:00 p.m. I smiled, picturing the look on Caveman's face. As I write this, the smile returns.

★ ★ ★ ★ ★

The Sunday after my escape, my wife, Cheryl, walked into the visiting room at Big Springs Prison Camp, with Dion, Dusty, Joshua, Lisé, Windy, and Israel in tow.

"Who are you here to see?" the guard at the desk asked.

"Don Ford."

The guard called on the radio. A few minutes later, another approached, taking Cheryl to the side.

"Haven't you heard?" he asked.

"What?"

"Your husband escaped last week."

Cheryl broke down and cried.

My smile becomes a frown.

FUGITIVES

Lee and I spent a couple of weeks in the basement of David McCasland's ranch house. Then David drove us to Fort Stockton, Texas, where we rented a room in a cheap motel. I called Oscar from the motel and made arrangements to go to Mexico. The next day Oscar picked us up and took us to Piedritas.

We spent a couple of weeks idle, wandering around Piedritas and the surrounding countryside. I offered to help—doing anything—but Oscar refused to let us work. The women worked constantly, the kids played and went to a tiny school. We befriended some of the local residents. Among them was one old, old man who had his own house but no means of support, other than what Celerino, Oscar's father, gave him. He smoked *Delicados** furiously and Lee joined him. He watched in amusement when Lee and I smoked a joint and laughed when I offered a hit to him. He had no interest whatsoever. His job was simply to watch—and watch he did—all day long, day after day, drinking coffee, smoking cigs and watching. He subsisted on eggs, tortillas and occasionally a bean or

---

* A notoriously strong nonfiltered Mexican cigarette.

two. On one occasion, I did see him eat a piece of goat meat, but it was difficult because he had very few teeth, none of which had an opposing tooth with which to connect.

Lee and I ate at Celerino's house. The Mexicans of Piedritas ate four meals a day, the first of which, the *desayuno,* was informal and consisted of coffee and perhaps a pastry. We took this meal on our own, if at all. Mid- to late morning they ate a large meal called *almuerzo. La comida* was served around two in the afternoon followed by a *siesta* and *la cena* late in the evening.

In between these events, Lee and I hiked, smoked, and watched. There existed a strong sense of community in Piedritas—in spite of the poverty, the people seemed content with their lot in life and looked out for each other far more than you'll find in the average American community. While they had very little, they did take pride in what they did have.

⋆ ⋆ ⋆ ⋆ ⋆

North of Piedritas, on the way to the cemetery, lay a field of junk—pieces of rusty metal, perhaps at one time parts of pickups, ancient appliances or who knows what. I heard sound coming from that place nearly every day, carried on the wind even though it was far from the town. One day I decided to investigate.

I had noticed a young man coming by Celerino's house routinely when he delivered water and mesquite firewood to Celerino's. Each time he came, he received some sort of handout, but Celerino was careful not to let us see what he gave the young man. This fellow grinned hugely and often laughed for no apparent reason. He seemed to be a nice kid, perhaps in his mid-teens. I learned he was fatherless and that his mother was very old. I was also told he was *not all there.*

I trailed this young man one morning without letting him know I was observing him. As he walked out of town, I watched from behind a huge boulder. He went to the field with all the old junk, where he had carefully arranged things. He took what looked like a piece of pipe and began striking the various pieces, each of which emitted a unique sound. As he struck the pieces, he also cried out toward the sky with sounds that originated from deep within. He altered the pace, beat and the intensity of his music, sometimes going soft and easy and at other times pounding almost maniacally. It was obvious that this whole thing was some sort of a prayer ritual. Some of the sounds he made reminded me of Native American chants.

I didn't find this funny in any way. The young man was dead serious. I don't remember ever hearing anyone pray more fervently. I couldn't help but wonder what went through his mind.

Later I asked Celerino about him. He just grinned and said nothing but made some sort of sign like the guy was loony. I'm not sure he was crazy. Maybe he was the sane one and the rest of us are nuts.

* * * * *

One of the more obvious shortcomings in the town of Piedritas for a couple of Americans was the lack of a sewer system, toilets and the amenities we expect to accompany them—like toilet paper. One of Oscar's workers, a man by the name of Beto, showed me which plant was prized for its ability to wipe an ass, a plant I believe to be called mullein. It has broad, velvety, light green leaves. While it may be okay—if absolutely necessary—it's far from being a nice roll of double-ply. It's much too easy to punch through the leaves. To the novice user, this usually meant ending up with more shit on your hand than on the velvety leaves.

Some locals carried small pocket-sized books around, with paper similar to what we call comic books. These would be read, or at least looked at, while shitting. That page would then be removed and used for wiping. This, too, left something to be desired.

Piedritas had an assortment of scroungy mange-eaten dogs whose bony ribs and lack of muscle indicated a poor diet. Every time I walked off into the sticks to take a dump, I noticed that one of these dogs would trail me. After I had attended to my business, he would attend to his—eating what I had so haphazardly left behind. This may seem gross, but I found it preferable to stepping in someone else's contribution.

* * * * *

After a few weeks, McCasland contacted Oscar and said he would be flying down to pick up a load. Oscar had acquired about 350 pounds he was willing to front to help us get started. Lee and I were installed in a line shack about five miles north of town with this marijuana. With nothing else to do, we decided to smoke a bunch of it to test the quality. I didn't want to sell a product I was not sure about, so we were diligent in testing to make sure it was okay. This sampling process was expected to last a couple of days or maybe a week at most. A month later found us still sitting, waiting and diligently doing our job.

It became apparent that McCasland wasn't going to show. I told Oscar to go ahead and sell the weed to someone else and he did, a little at a time. Normally

Oscar came alone, picked up what he needed and delivered the marijuana to the river, but one night a young white American man and his woman showed up at our shack to pick up their load. They looked at us curiously, startled to discover two Americans—with hacking coughs—camped out in the middle of the Chihuahuan desert, guarding marijuana, both of them beginning to look like Grizzly Adams.

Now what were we going to do? We had arrived in Piedritas with all of ten dollars between us—escaped fugitives. We couldn't go back to the United States. We were worthless as overland smugglers and we had no plane. Oscar wouldn't let us work. Neither one of us was well suited to idleness. We wanted something to do. We came up with a proposition for Oscar.

"Why don't you let us grow some marijuana for you?" I asked him. "Those guys you work for end up with all the money. Why don't you grow some here?"

"We tried and it didn't do well," he explained.

I was a marijuana grower from way back and I had learned quite a bit about farming in a desert climate when I managed my dad's farm near the Pecos. "I can grow it if you give me a place with a supply of water and a handful of seeds. We'll do it for forty percent of the wholesale value down here and you can have the remaining sixty percent for selling it."

Truth was, I had already accumulated the handful of seeds from my job as a marijuana sampler. Several hands full.

Oscar assured me he could get permission from those in control of the drug trade to grow a patch of pot. I had doubts about that because Oscar had told me a few times before that everything would be okay and then it wasn't. But I figured it was worth a shot. He had doubts about my ability to grow anything in the harsh desert climate, but giving us the chance to try would make his life easier. He could put us out in the sticks where we wouldn't be seen. Then he wouldn't have to explain what two gringos were doing living in the town of Piedritas.

Oscar moved us out to another *jacal*, or line shack, five or six miles south of Piedritas and we went to work.

* * * * *

The line shack was built into the side of a small rock-covered hill. It was very hard to see. The original builder had dug into the side of a rock outcropping and used the large rocks he excavated to form two sides and a front wall. The back wall was the natural rock face of the hill. The side walls were partially composed of the same, except that as you moved forward, the ground sloped down and the stacked rocks maintained the altitude of the back wall. The front wall was made entirely of stacked rock.

A sheet of thick tin served as the front door. No mortar held the rocks in place. The entire shack was maybe twelve feet wide by twelve feet deep with barely enough headroom for a man of my size. The roof consisted of lashed tree branches with a foot-thick layer of spent *candelilla* plants on top. It provided shade but was not impervious to water—rains, however, are rare in that part of the world. The structure did a good job of keeping most of the hostile environment outside, excepting an occasional rattlesnake or a scorpion that found its way through the cracks. Two small metal bunks were installed near each wall for us to sleep on. Smoke from the fire escaped through a gap between the walls and the roof and also filtered through the *candelilla.* Previous campfires had left their mark on the back wall.

The house had originally been constructed next to a candelilla camp. There was a good hand-dug water well about a hundred yards away and an ash-filled trench where the *paila* had been. At some point, the camp had been abandoned and a *curandera* (female witch doctor) had moved in and set up shop. Lee and I found circles of colored rocks and other articles in various places surrounding the house. Both of us avoided messing up these monuments out of respect for the woman and her beliefs—and, also, just in case there was something to those beliefs. We sure as hell didn't want any curses visited upon us.

Oscar left us a few basic necessities: a pot to cook in, a sack of beans, potatoes, rice, a sack of flour, lard, baking powder, eggs, tomatoes, peppers, and coffee. He also left Lee a supply of cigarettes—Lee would buck quick without coffee or cigarettes. The metal lid off of a fifty-five-gallon barrel served as the *comal* on which I made tortillas. He installed a large black plastic barrel in one corner and filled it with water brought out in the back of a pickup. We were given a few buckets and some rope, a grubbing hoe and a garden hoe. Oscar also provided one more thing at my request—about a quarter pound of marijuana to smoke. I would buck quick without that.

One of our first tasks was to clean out the well. It hadn't been used for some time and had collected a thick layer of sediment and algae. The well was six or eight feet across (eight or six, if you happen to be Mexican), lined with rock without mortar and was ten feet deep. The water came up to five feet from the surface. Lee and I found an old homemade ladder and put it into the well. I went down into the water and scooped buckets of mud from the bottom while Lee worked above, pulling these heavy mud-and-water-laden buckets to the top with a rope attached to the bucket. At times we switched places. I was a relatively strong young man, so I was surprised to see how well Lee held up to the arduous work. Over the next few months, he would prove time and again how tough and resilient he was for a man his age—for a man of any age. He worked in nothing but a pair of shorts. His skin became as brown as most dark-skinned Mexicans. I am the son of a redhead and cursed with an inability to tan, so I tried to stay covered but managed to get sunburned in spite of my best efforts.

Lee and I prepared a small starter bed with a grubbing hoe. We planted thousands of marijuana seeds, covered them with dirt and hand carried five-gallon buckets of water to wet the soil. A few days later, young plants emerged and stretched out tiny leaves to gather the rays of the sun.

Lee and I then began breaking the rich, hard-packed soil which was to become our field. Each stroke of the hoe would break loose a piece of ground about four inches wide and six inches deep. Lee would swing right-handed for a time and then switch. When he tired, I would take over.

We made small round dirt borders in the now-loosened sandy loam soil and filled them with water, once again using buckets and ropes to extract water from the well. Once the water was absorbed, we transplanted between three to five young plants from the starter bed into each hole in the field. It took almost a month of steady hard labor to get the ground prepared and the field planted using this crude method. But, by the time we finished, we had over an acre—maybe closer to two acres—planted. As the garden grew, watering became an all-day task, interspersed with hoeing the earth around the plants to keep the soil loose and to conserve moisture.

In a desert climate, with a plant like marijuana that produces a taproot, heavy soakings are preferable to daily watering. A time will come when the extreme summer heat will bake the plants. In order to survive, they'll need strong deep root systems. Daily watering causes the plant to develop shallow roots. To promote good roots you have to soak the ground thoroughly and allow the surface to dry between waterings. As the plant is stressed, it sends roots down into the ground in search of moisture. The sun and heat cause the ground to crack. These cracks allow moisture to escape. Rather than water the ground when these cracks form, a farmer should thoroughly loosen the top of the soil. This loose soil fills these cracks. Contrary to what you would think, it traps the moisture deep below, and of course, destroys competing weeds. The young plant must be allowed to suffer some heat-related stress to develop toughness. I suspect that's why Oscar had failed in his earlier attempts to grow marijuana.

Marijuana, unlike most plants which have both male and female sex organs on each plant, has male and female plants. For smoking purposes, the female is the most desirable of the two—by far. The female plant can be made even more desirable by removing all male plants from the garden and thereby interrupting the breeding cycle. The resulting product is known as *sin semilla,* or seedless marijuana. For that reason, we started between three and five plants in each hole. The plan called for removing all the male plants once they showed their sex. We didn't want to leave unused space in the garden.

The plants grew and the task of hand carrying water became more and more difficult. Oscar showed mercy on us and brought a gasoline-powered centrifugal pump out, along with black plastic pipe and a flexible hose we used for watering. Lee and I made deep borders around three or four rows of plants, eliminating the individual hand-dug circular borders. These new *melgas* were surrounded with earthen borders at least a foot deep that allowed us to thoroughly soak the ground with much less effort. The plants were several feet high by this time. At about this stage, the plants began to show the first indications of their sex. As they did, we removed the males. When more than one healthy female plant remained in an individual hole, we extracted the excess plants and transplanted them to a new area, further expanding the garden.

Oscar began to show interest in what we were doing when it became apparent we were going to be able to produce a crop. His supply of marijuana from down south dried up as it nearly always does during the summer, so Lee and I dried and smoked leaves, particularly the leaf bud tips from these plants. At this stage of growth, it was of poor quality but better than nothing. We removed the central growing tips from the female plants to cause the plants to branch out rather than growing straight up and we consumed these also.

Blisters turned into callouses and our bodies became hard from the intense daily exercise. I did all the cooking while Lee did the dishwashing. He was very particular about it. He hated soapy aftertaste and would rinse the dishes at least twice before he was convinced the job was properly done. Lee also washed our clothes by hand and was very thorough in the way he did this job as well. I was more than happy with the arrangement because I liked to eat and cook and was not crazy about washing dishes. The haphazard way I did the job, the time or two I did attempt it, convinced Lee that I was not suited for that kind of work.

Either Oscar or one of his hands would come by once a week to provide us with our weekly ration of food and material. One day Oscar showed up and told us we had visitors.

* * * * *

When I was in Big Springs prison, I had given Cheryl a phone number. I told her that if anything happened to me, she was to go to a pay phone and call the number. When I escaped, she did as I had told her. Jimmy Hobbs told her where I was. She was then able to get through to Oscar and arranged to meet us at La Pantera. She planned to bring clothing and supplies. My cousin Risa also came along.

They came one evening and camped out on the American side. Lee and I waded the river and joined them. It had been some time since I had been with a woman so I did little sleeping. Months later, I would learn this encounter was responsible for the conception of our last son: Levi Winston Ford. Sometimes it's strange how fate works. If I hadn't escaped, he wouldn't be alive.

Cheryl told me U.S. Marshals had questioned her concerning my escape. I told her to go home and file for a divorce so they'd leave her alone.

The next morning we sat talking and drinking coffee around an illegal campfire when several horse-mounted park rangers appeared. Lee and I saw them coming while they were still about fifty yards away. We wasted no time saying goodbye. Both of us turned and ran for the nearest stand of salt cedar and brush. I can run fast, but Lee matched me stride for stride. The rangers followed but the brush was too dense for a horse to enter, and I suppose they were afraid to follow, thinking we might be armed and waiting to ambush them. I don't think they knew who we were and most probably assumed we were Mexicans.

I found a depression in the ground, buried myself with natural debris and lay there quietly. I heard transmissions coming from one of the ranger's radios—a time or two it sounded like he was no farther than twenty feet away from where I lay. I was reminded of times when I played hide-and-seek as a kid, a favorite game of ours. I was conscious of the sounds of my breathing. A stick dug into my side but I didn't dare move. I needed to piss. I had no idea where Lee was.

After about thirty minutes, I heard the rangers talking to Cheryl and Risa. I snuck to a spot where I could observe. It didn't appear that he was going to arrest them so I continued down to the river and crossed on foot. I began to walk toward a line shack I knew about, some five miles or so from the river and ran into Lee. He, too, had escaped, using the same tactic.

Risa and Cheryl were cited for the campfire and allowed to leave.

* * * * *

Cheryl had brought me a small but razor-sharp hunting knife and a whetstone. I never went anywhere without a knife. I still don't. I feel naked without one.

I took my prized possession back to Piedritas. Since I didn't have a lot to do at this particular phase of our stay, I spent time seeing just how sharp I could get this knife. I sharpened and sharpened, shaving the hair on my arm occasionally to see how it was going.

Sergio, one of Oscar's little brothers, saw how sharp this knife was and asked if he could borrow it to kill a goat for the evening meal.

"Claro que sí," I replied, glad for it to be of some use.

We followed him to Celerino's house.

"Where's the goat?" I asked.

Sergio led me to a small house behind the main house and opened the door. I thought it strange that the goat would be in a house. He walked in and looked around, seeing nothing. Finally, he looked under the bed. Seconds later he dragged a tiny kid goat from under the bed. The goat cried out.

I stared in horror. "That's just a baby!"

"They're better that way," he told me.

"But wouldn't you get more meat if you let it grow a little more? He still sucks, no?"

"Yeah, but they are better this way. His *tripas* will be clean."

I'm a meat eater. I have killed animals to eat their flesh. For some reason though, the thought of killing baby animals to eat just seems cruel. At least an adult animal has had some time to experience life.

Sergio dragged the goat out into the yard and ordered a hand to bring a pot from the kitchen.

The man returned with the pot. Reluctantly, I handed Sergio my knife. The goat stood there bleating. Sergio sliced his neck. The goat screamed. Sergio held him in place as his blood flowed into the waiting pot. Unconsciousness arrived quickly for the baby goat but that minute had to have been terrifying for the little guy and for me as well. His body relaxed as they drained the last drop of blood his body would yield into the pot.

"What are you doing with the blood?" I asked.

"We're going to eat it," Sergio replied.

We, my ass, I thought.

Sergio strung up the goat by his hind legs and began the butchering process. He carefully cut along the inside of both hind legs, from the anus to the hocks, cutting from the inside of the skin rather than from the outside to keep from cutting into the hair and also to prevent cutting into the flesh of the leg and contaminating the meat. He then cut a circle around the anus, separating the digestive tract from its external connection to the animal.

Next he cut a line straight down the middle of the goat's abdomen, careful not to cut into the gut liner. This incision was also made from the inside out. At the chest region, he cut up each arm in the same way. Then he went back up to the hock region of the goat and cut a ring around the legs. This gave him the starting place to remove the hide. Beginning at the hocks, he peeled the hide downward. Once the hide was peeled all the way down to the goat's head, he severed the head from the body by cutting all the way through the neck. The remaining hide was then peeled to the knee joint of each front leg and then the legs cut off by severing the connective tissue in the knee.

Sergio rinsed the goat's body to remove any blood or hair that might have remained and then cut the gut liner, careful not to rupture the intestines. He then reached into the body cavity and retrieved the end of the colon from the confines of the pelvic bones. He pulled this out until the small intestines were revealed. These he cut into pieces, removing the milky looking material inside.

"What are you going to do with that?" I asked.

"You'll see."

"You eat the intestines?" I asked.

"Sure."

I never had tried intestine to this point in my life; the thought didn't appeal much to me. "Aren't they dirty?" I asked.

"Not in a baby. It's just milk," he added, squeezing more of the white material out of a sectiion.

After collecting all the intestines, he began to harvest the rest of the organs. Nothing was wasted. Part was put into one pile along with the intestines and others were cut up and dropped into the blood. Once done, everything was sent to the house to be cooked by the women.

That evening, Lee and I sat down to the evening meal, exchanging worried looks. A skillet was placed in front of us. Oscar, Sergio and Celerino looked on as Lee and I spooned small portions of the unidentifiable body parts onto our plates. *Thank God for the refried beans and tortillas,* I thought. I knew I had to eat some of this stuff to keep from offending my hosts. Tentatively, I looked through the contents on my plate. I tried a piece that looked like liver. It almost came back up. I don't eat liver. I suppose if I were starving to death I would but the dog has to get something, doesn't he?

Chew, chew, chew, now swallow—hold it, hold it—now another bite.

I removed a piece of something resembling a noodle, which I knew was a piece of intestine. I ate it. And guess what? It wasn't bad. I then went after all the intestines on my plate but there was no way I could get down all the liver. I'd guess it must have been pieces of the heart that also seemed tolerable. When done, a portion remained on my plate. I apologized. I guess that was better than throwing up at the table.

The next day, the congealed blood and parts were cooked into a loaf of sorts. I came up with some excuse for missing this meal and went to bed hungry—and thankful to be so.

LIFE IN THE BADLANDS

All my life, I thought that the white American cowboy was a superior stockman to any cowboy in the rest of the world, certainly to any Mexican. A skinny little Mexican hand by the name of Beto changed my mind.

Beto stood about five-foot-six, and probably weighed about 130 pounds. His skin was as white as mine. His hair had a reddish hue to it and his eyes were green. Beto didn't go anywhere without a cowboy hat and a long-sleeved shirt. Most of the time, he was the one who brought out our rations. When he did, he arrived on horseback. His primary responsibility, however, was the growing herd of cattle Oscar ran on the *ejido's* land, as well as a band of about thirty quarterhorse broodmares, which ran loose in the countryside.

Oscar set up a drinking trough for cattle near our shack. One of our responsibilities was to run water from the well through a stretch of inch-and-a-half black plastic roll pipe to fill the trough. There was no other source of drinking water for miles. Because of this, the land wasn't being grazed. Oscar's men pushed some cattle out there. Then one day he showed up with two young imported Limousin bulls he bought in Texas.*

---

* A Limousin is a large breed of beef cattle, imported from France and relatively new to the Americas. They are very large, naturally polled (hornless) and usually a sandy red color.

We didn't have any wire to keep the cattle out of the marijuana patch, so Beto showed us how to construct a corral using thorny brush to form a barrier. He caught a milk cow for us by first catching her calf, then using the calf to lure her into a makeshift corral. From that day forward, we had a supply of fresh milk.

When these two Limousin bulls were dumped out in the middle of this desert, one of them decided he wanted no part of the place and turned north—heading for home. The following day, Beto arrived, looking for the bull. I pointed in the direction I had seen him run. Beto had a brown nylon rope with him that must have been three-quarters-of-an-inch thick and sixty feet long. He rode a big stout bay gelding. This bull he was after weighed in the vicinity of two thousand pounds and had no respect for man or horse.

"What're you going to do?" I asked.

"I'm going to bring him back," he replied.

"How do you plan to do that?"

"I'm going to rope him," he said. Beto was shy and soft-spoken; he seldom wasted any words.

Yeah right, I thought to myself.

He found the bull's tracks and started after him. Three days later, he showed up dragging the bull along behind. I couldn't believe my eyes.

"Where did you find him?" I asked.

"In the park. He swam the river."

"You rode all the way to the park and brought him back?"

"Sí."

"How did you get him back?"

"I roped him."

"You roped that bull in the park and drug him all the way back here?"

"Sí."

Beto had tracked the bull thirty miles to the river, crossed into the United States, found the bull—a huge feat—and then roped the crazy bastard, who outweighed his horse by eight hundred pounds. Then he crossed the Rio Grande and stayed tied to him for thirty-six straight hours, all the while keeping himself and his horse alive. The bull was dead broke to lead by the time he got back. Beto told me the bull had charged him at first, but he tied him to several trees along the way to teach him a lesson. The bull stayed put after that. You won't find ten white cowboys in the United States of America that could have pulled off that stunt. You might not find one.

Another time, a Mexican cattle buyer showed up, wanting to buy two big calves. Beto and another young cowboy rode up with their horses. Celerino,

Oscar's daddy, pointed out which calves he wanted caught and these two men unfurled their ropes.

This I got to see, I thought to myself.

Their ropes were very long, at least sixty feet. The two men started toward the first calf and he took off. The young cowboy twirled his rope and ran after the big calf, while Beto hazed the animal.

"No way," I told Lee.

The calf was on a dead run and turning when the Mexican let sail with the rope. It arced skyward and then began to fall—as it fell, the size of the loop diminished until it was barely any larger than the calf's head when it arrived and found its target.

"Lucky shot!" I exclaimed. That calf was over forty feet from the horse where he caught him, and both the horse and the calf were on a dead run.

"He'll never do that again in a hundred years," I told Lee.

They dragged the calf up and into the back of the waiting truck, which had been equipped to haul livestock, and then went after the second calf.

My jaw almost hit the ground when the scene was repeated in almost identical fashion. I later discovered that in some rural areas, while we're learning to throw and catch a baseball, these young men are learning to use a rope. And learn they do.

* * * * *

Beto was also a good campfire cook. He showed me how to improve my skill at making tortillas and refried beans. When this is all you have to eat, you want them to taste good. Believe it or not, five cooks can start with the same ingredients, and each one will come out with a slightly different product in size, texture and shape. The women in that region all make their tortillas small, and the men make them large—why I don't know. Sometimes Beto would bring some ready-made tortillas for us from someone in town. After awhile I could tell who made them, almost as though they carried signatures.*

* People around Piedritas ate more flour tortillas than corn, but this wasn't always the case and still is not the rule in other parts of Mexico. What corn tortillas they do eat are made from *masa harina*, an instant powdered form of cornmeal that can be mixed with water, formed into balls and pressed in a device between two pieces of cellophane to make the tortillas. Long ago, corn was readily available. Wheat was not, so people soaked the dried kernels of corn in limewater until the shell of the kernel was dissolved, leaving a product similar to what we call hominy. The hominy would then be taken to a local mill and ground wet, after a thorough rinsing—the resulting product was referred to as nixtomal. Old timers swear the tortillas made like this were far better.

Beto first boiled his beans in a covered pot. He would never add cold water to the boiling beans. He kept a second pot full of plain hot water alongside the bean pot to replenish water as needed. No seasoning was added, other than maybe a small amount of salt. But even this wasn't necessary as it can be added later. Once the beans were relatively soft but the broth still thin and watery, he heated a skillet and added a large dollop of lard. The lard was allowed to melt and get very hot, almost smoking. He then dipped the beans from the pot, leaving all the broth behind, and added them to the hot lard. The lard scorched the beans and shriveled the skins as they fried. He told me if you try to do too many at a time, the lard will cool too fast to properly fry the beans. He added salt at this point—and removed the skillet from the fire. Then he mashed the beans using the end of a thick dowel rod or a section of handle from a tool. After a thorough mashing, he added broth from the bean pot and mashed them more, and then stirred this concoction. This was repeated several times until the beans were very thin and watery. He then replaced the skillet on the fire and allowed the beans to cook until thickened, stirring occasionally. Once they were close to ready, he would set them farther away from the heat to simmer slowly. A plate of Beto's beans was always offered with a smile—the smile of one who cares. Caring is the secret ingredient of any great cook.

You'd think a man would tire of eating almost the same thing every day, but Lee and I never did—though if I had to eat what we call refried beans and tortillas here in the United States, I'd be sick of them in two days. The only meat we ate was an occasional piece of goat Beto brought out. Without refrigeration, it made little sense to slaughter a beef—besides, they were worth too much money. A time or two we killed quail with a cheap .22 rifle. Once in a while we ate a cottontail rabbit, although I was a little afraid of them. I'd heard of people catching Rocky Mountain spotted fever from eating rabbits. I also was leery of undulant fever from the unpasteurized cow milk, but I drank it anyway.

I put in an order for *piloncillo,* a crude, cone-shaped type of raw brown sugar. Using flour, milk, eggs and baking powder, I came up with some pretty decent pancakes. The *piloncillo* made good brown sugar syrup when melted with water and allowed to thicken in a pot over the fire. We even went so far as to make butter a time or two. Another change of pace was *sopapillas,* which are very easy to make. I made the dough for *sopapillas* just like that for tortillas, except I didn't add lard or shortening. The thin rolled dough was placed in a hot skillet full of lard or shortening and fried. When turned at the proper time, they puffed up, leaving a crispy exterior and a hollow center. A dusting of brown sugar and cinnamon made them better. If that wasn't available, the *piloncillo* syrup or honey also worked well. Many nights we went to bed full of *sopapillas* after smoking plenty of marijuana and drinking coffee. Prior to knowing Lee, I rarely drank coffee, but after living with him for a long time, I drank the stuff constantly. Several months later, I learned that caffeine is truly addictive when I was forced to quit the stuff, cold turkey.

★★★★★

Without anything to read, movies or TV to watch, Lee and I entertained each other by telling stories. I told him all about my childhood and my family. He was particularly interested in learning about my father because they were roughly the same age and shared some of the same interests. My dad had served in the military during the Korean conflict in the counterespionage branch of the Army. Lee had gotten in on the end of WWII. He had lied about his age and got away with it. He served in the Navy. Later he got out and joined the Army. The Army taught him how to fly—P-51 Mustangs.

Lee did well as a pilot, but often got in trouble because of his playful nature. He was known for buzzing the tower during training exercises and many other disobedient acts—offsetting his abilities as a flyer.

When his military career ended, he began flying as a crop duster. That's how he met his first wife. She worked as a crop duster, too. Lee said she was the woman of his dreams. He loved her dearly. He told me they had never had an argument, but their marriage was ill-fated. Thirty days after they were married, they flew off together into a storm, each in a separate plane. Lee lost track of her. She was never seen again—nor was her plane. Lee believed she became disoriented, flew out into the Gulf of Mexico and crashed into the ocean.

Lee remarried, but didn't get along well with his second wife. They had five children, but he was unhappy at home. He learned that a guy by the name of Castro wanted to buy airplanes for a revolution going on in Cuba, and decided he would take him one. He made the mistake of stopping in Florida to do a little drinking—and perhaps some whoring around—and told the wrong person his plan. His wealthy father-in-law was also his boss at the time. When Lee got arrested, he arranged for Lee to go to a psychiatric hospital instead of doing time as a criminal. To get out, all Lee had to do was prove that he wasn't crazy. Lee figured this wouldn't be a problem—after all, he wasn't crazy.

He went to this institution and began trying to prove his point. After six months of fruitless effort, a nurse finally pulled him aside and told him, "You should have come in here and bounced off of the walls for a few days, and then acted like you got better. You came in here acting normal. These guys are going to study you forever, trying to figure out what's wrong with you."

After receiving this advice, Lee was able to con his way out of the place, but his marriage failed. He continued flying, at one time befriending Janet Guthrie, the first woman to drive in the Indianapolis 500. Lee helped teach her how to fly at a very early age. He never said so, but I suspect they were also lovers.

Lee began smoking marijuana at the age of twelve. When he was in his forties, he got caught in Texas with a small amount. Texas was not the place to get caught with marijuana during the sixties. He served seven years for possession of a couple of ounces.

Lee had been an inmate within the Texas Department of Corrections. He told stories about working the fields and hoeing all day long. Field bosses punished disobedient inmates by dragging them out of the fields with their hands cuffed over the saddle horn of a horse. They would have to shell peanuts in the hole all night long and then work the following day without having slept or eaten. After two or three days of this, even the most difficult inmate kept up or got shot. Building tenders—inmates given preferential treatment—kept order in the prison and did the dirty work when the guards needed to punish someone.

We had nothing to read but Lee quoted books to me from memory—books like *Catch-22* and the works of Hemingway, Steinbeck, Orwell, Clavell, Vonnegut, and many others. He loved science. Though he didn't have a degree, he knew an awful lot about geology, which happens to be my father's field of expertise. He pointed out the various layers of earth visible on surrounding hills and described the prehistoric times when they were created. Lee described his travels around the world as a merchant seaman. He had seen places like Russia, India, Japan, Europe, the Philippines and Indonesia first hand. When he was there, he'd talked politics with the people, eaten their food and slept with their women.

Lee considered himself an agnostic. I professed faith in God and Jesus. We spent many lively evenings debating our views. He made me question some of my beliefs and I think I also challenged some of his. In spite of Lee's lack of religious belief, he did operate under a strict moral code, perhaps an unusual one, but I was a witness to the way he lived. He had love in his heart for others less fortunate in life than himself, a lesson many proclaiming to be Christians need to learn.

TROUBLE IN PARADISE

One day, to my astonishment, Cheryl; Lee's wife, Glenda; Glenda's daughter, Stephanie; and Dion, my oldest son; crossed the international bridge at Eagle Pass, Texas, and found their way to Piedritas. They spent one night. Cheryl told me she wanted to move to Piedritas. She also wanted to bring the rest of our children. Glenda asked permission from Lee to do the same. I tried in vain to convince Cheryl to wait until after the marijuana was harvested, but she insisted.

Her argument was simple: *She was my wife and our children were our joint responsibility.*

I hoped Oscar would trump her, but when I asked him what he thought, he said it would be okay. If he didn't object, then who was I to say no?

Glenda drank a lot. When she asked if she could move to Piedritas, I caved in and told her she could but only if she would stop drinking. Cheryl and Glenda went back to Texas, gathered supplies and returned with the rest of my kids in tow. During my stay in Big Spring, Cheryl had totaled my van. She returned in a used Ford pickup my father bought for her. She also brought several tents. I rented a house for her in the village, but Cheryl rarely wanted to stay there, insisting she wanted to be with me. So the whole crew moved out to our little *jacal* and the surrounding countryside and lived like desert nomads.

We had six children by this time, and Cheryl was pregnant with the seventh, conceived at La Pantera during her first visit. So, there we were—a small community of white folks growing marijuana and raising children in the Mexican desert.

Glenda went through a nightmarish withdrawal, complete with deliriums, sweating, convulsions, nausea, tremors and hallucinations, declining in severity as time went on but lasting a full month before they were over. I was shocked at the severity of her withdrawal. There were good days when she appeared normal. Other times it looked as though she might die.

I did appreciate getting to be around my children again. Not all my memories of this time are bad. I had never been around Israel, our youngest, so I spent some time getting to know him. Domestic duties—things like cooking, washing clothes and dishes—increased with the additional people. Cheryl had camped out before, but Glenda had a lot to learn about life without its modern conveniences. As the patch continued to grow, I convinced Cheryl to stay in town more often, but she routinely came out to visit and spend the night anyway. On rare occasions, I went to town to stay with her.

One day I watched Dion, Dusty, Joshua and Stephanie as they played. My head ached slightly and I felt groggy and depressed. I had run out of good marijuana to smoke and had neglected drying any leaves—which weren't all that good anyway. The kids laughed and ran and joked, without a care in the world. The sounds reached through my gloomy state of mind.

I can remember being carefree like that. How did I get like this, where I can't be happy without drugging myself?

I took a look at myself and didn't like what I saw.

Then one day I saw Joshua put something into his pocket. I searched his jeans and removed a baggie full of dried marijuana leaves. The bag also contained rolling papers, matches and a crudely fashioned joint. He was seven years old.*

One morning, I came running when I heard Cheryl crying. One of the two Limousin bulls had broken through the thorn barrier around the marijuana patch and was busy breaking down our marijuana plants, which by this time were about ten feet high. Both Lee and I sprinted toward the patch. Lee carried a single-shot .410 shotgun, but I had nothing. We got in the patch and began to try to scare the bull out of there. He wasn't impressed.

The bull squared off at Lee. Lee waved the gun at him and shouted. The

* When he is offered marijuana now, my oldest son, Dion, laughs and says, "No thanks, I quit when I was twelve."

bull just stared. Finally, he charged. Rather than shoot, Lee tried to hit him with the barrel of the shotgun. The bull felt nothing and planted his massive head right into Lee's abdomen. Lee was no match for two thousand pounds of muscle, sinew, bone and attitude. The blow pushed him backward and down. He landed flat on his back, without the shotgun. The bull rammed his head into Lee's midsection, coming down on him with the full force of all that weight. Fortunately for Lee, the bull was hornless. I picked up a softball-sized rock, which I threw as hard as I could into the bull's rib-section from about five feet away. It had no effect whatsoever. He continued to crush Lee. I ran up to his head and slapped him and yelled. Finally, he turned his attention to me. As he charged, I did a bullfighter move at the last possible second, jumping out of the way as he ran by, head down and charging.

Lee scrambled back to his feet and grabbed the shotgun. The bull looked at him again from ten feet away. Just as he was about to charge, Lee shot him in the nose. The bull was startled by the shot. He raised his head and took a step back. Blood flowed from his nose, but the shell Lee used to shoot him was birdshot, and even at that close range, had done no serious damage. He looked again at Lee, who now stood with an empty weapon. Glenda ran to the edge of the field with a twenty-gauge pump shotgun. I took it from her. The bull refused to leave. I shot him in the ribs a total of three times before he jumped the barrier and got out of there.

Lee ended up with a few broken ribs out of the deal, but the bull would be tormented for some time. Nearly every time Lee saw him from that day forward, he shot him in the ribs with that twenty-gauge. It never did kill the bastard.

* * * * *

One day, Oscar showed up with a man he introduced to us as Alejandro Cerna. Cerna was a Colombian national, short in stature, with a round build like the actor Danny Devito. He was interested in flying large planeloads of marijuana and cocaine from Colombia, landing in northern Mexico and redistributing the large loads into smaller ones, using the traditional routes Mexicans used to get their products into the United States. He wanted to talk to Lee about setting up some sort of a navigational device for landing a DC-6. The proposed airstrip was an unpaved clearing near the tiny town of San Miguel, Coahuila, about five miles south of our *jacal*—the exact middle of nowhere. Lee agreed to help the man.

A few days later, a small Cessna 172 buzzed our pot patch. Several hours later, David McCasland showed up in Oscar's company. McCasland came up with some lame excuse for having abandoned us in Mexico after our escape. But it didn't really matter. He was there and ready to do business, and Lee in particular was excited about it. Flying was much more up Lee's alley than mine. He had no desire to be a farmer or to spend the rest of his life stuck in Piedritas. David and Lee talked about flying loads of the Colombian marijuana back to David's place near Tucumcari, New Mexico. David had arranged to buy a Cessna 206 turbo airplane. Cerna offered him $50,000 for each 800-pound load he could get back to the United States. Cerna's people would take over once the merchandise reached American soil. I was not involved in this, other than acting as a translator. I didn't want to be. I was not a pilot, and our patch was getting large by this time. I hoped it would yield a minimum of several thousand pounds of high-quality *sin semilla,* which would be worth a half million dollars at Mexican prices. It had the potential to be worth far more, maybe as much as two million. My forty percent share would be more than enough to set me up with a home and a legal business in Mexico.

The huge DC-6 from Colombia landed on a flat piece of dirt near San Miguel. It took a small semi-truck, a large gathering of men and a caravan of pickups to unload the marijuana, which was taken to abandoned fluorite mines northwest of Piedritas. The entire time the marijuana and cocaine were being offloaded, fuel was being pumped into the plane for the return trip. Forty-five minutes after it landed, it was gone, headed back to Colombia.

McCasland bought the Cessna 206 turbo, flew to Piedritas and hauled a load home, paying for the plane in one trip. He made plans to buy another airplane so Lee could join in on the action. Another pilot connected to Cerna began to fly out loads as well.

Another group began to smuggle the marijuana concealed in the tubes of large twelve-man life rafts. Justin Adams, a guy from Georgia, headed this operation. It was common for American tourists to raft the Rio Grande. Three large and beautiful canyons separate our countries in the Big Bend region. Justin launched three life rafts into the river at La Pantera. In them rode a crew of good-looking scantily dressed girls and a few guys, ice chests full of beer, and fishing gear—a perfectly normal looking group of tourists. They floated downstream to a place inaccessible from the American side, where they encountered Oscar and his crew. They beached the rafts in thick stands of river cane. During the night, they inserted 800 pounds of marijuana into the tube forming the body of each raft through a hole where the oarlocks rested. The patch holding the oarlock in place was then reattached with a device that melted the rubber, creating a watertight, smell-proof seal, and then the boat and its cargo got back into the river. The passengers floated to a private ranch near La Linda, got the rafts from the river, loaded them on top of vans and drove off. Each successful trip introduced 2,400 pounds of marijuana—with a wholesale value of $1.2 million—into the United States.

What none of us knew at the time was that the guy flying the DC-6, a pilot by the name of Michael Palmer, was being paid by both sides of the War on Drugs, namely Alejandro Cerna on one hand and the DEA on the other. He also worked for the CIA. I wouldn't discover this until years later when I was reading the March 1989 issue of *Playboy* while imprisoned in Texarkana, Texas. What I read represented just a tiny fraction of what these guys were doing at the time.

* * * * *

Oscar began to spend more time away from Piedritas and the farm. His companions partied and snorted a lot of cocaine. I suspect he did the same when he was with them. His brother Sergio went off the deep end on the stuff. Sergio visited us occasionally. His nose was always red and inflamed and he sniffed constantly. Every few minutes he would do another bump. He tried to conceal this from his father, but his condition deteriorated quickly and it became obvious to everyone that he had become an addict. Oscar never allowed us to see where he kept the cocaine, but it was there—somewhere.

Oscar seemed to lose interest in growing marijuana once the Colombian operation got into gear, but one day he showed up with a well-dressed man he introduced as a minister of the Mexican government. This man told me he was in charge of the *federales* for all that part of Mexico and worked out of Nuevo Leon. He checked out our marijuana patch, sampled some of the early product of the field and liked what he saw and smoked. Both he and Oscar convinced me everything was okay—the proper palms had all been greased. The man left us a couple of six packs of beer. I watched suspiciously as Glenda drank hers.

Right after that, Glenda and Cheryl convinced me they needed to go to Músquiz, Coahuila, to buy some things not available in Piedritas. I agreed to let them go, but before they left, I pulled Glenda aside.

"You and I both know you have a problem with alcohol," I told her. "If you go down there and start drinking, you might make a mistake. And if the cops get their hands on you, you'll jeopardize everything we have. Do you understand me?"

"I wouldn't tell them anything."

"Oh, yes, you would."

"I would not. I've been questioned by cops lots of times. Ask Lee. I never told them anything."

"This isn't the United States. A Mexican cop gets ahold of your ass, you'll talk."

"No, I won't."

"Glenda, if Mexican cops get you, you'll turn in your own mother for a crime she didn't commit before they get through. Don't fuck up!"

"Don't worry. I won't."

Before they left, I talked to Cheryl about the trip. They would be taking all of the children and were to stay in a motel. I told her to keep an eye on Glenda and not to allow her to get near a bottle of booze. She agreed. And off they went.

Cheryl showed up several days later with all of the children, including Glenda's daughter Stephanie, and a pickup load of merchandise—but without Glenda.

"Where's Glenda?" Lee asked.

Stephanie ran into the arms of her stepdad. I stood glaring.

"I don't know. Glenda went to a bar and began drinking. I couldn't make her stop, so I took Stephanie and went back to the motel. Glenda never showed up. I didn't know what to do, so we loaded up and came home."

Lee looked at me and apologized.

I felt terrible for him, and I was afraid for Glenda. I was also afraid for the rest of us.

A few days later, we found out what happened to her from XREY, a powerful radio station out of Monterrey. Glenda had been found in a ditch beside the road. She'd been run over by a pickup, but was still alive. Apparently she had gone out drinking with some men. After getting thoroughly sloshed, she got out to pee behind the pickup, which happened to be stopped on a hill. She squatted behind the truck. The driver accidentally allowed the vehicle to roll backward over her. After realizing what he had done, he drove off, abandoning her alongside the road. The next morning, she was found by a passerby, picked up and taken to a hospital. They removed her spleen and fixed a broken leg. They also looked in her purse and found a small amount of freshly cured marijuana. Although she was allowed to stay in the hospital until her wounds healed, she was placed under arrest.

* * * * *

In light of what had happened, I asked Oscar for permission to cut down the marijuana patch. The marijuana was not entirely ripe, which would mean a slightly lower-quality product and a smaller yield, but I estimated it would still make quite a lot of a decent smoke, more than enough to take care of our needs. But you can't move a patch of living plants. I knew we would be vulnerable once word of what we were doing got out. Nonetheless, Oscar told me to wait.

In spite of Oscar's instructions, Lee and I began to harvest individual buds and dry them, employing two local men from town to help, one a teenager, the other an old man by the name of Eliezar. As the buds were dried and sacked, we turned them over to Oscar to store or sell. What we cut was a tiny fraction of what was there. In all we may have harvested several hundred pounds, which would not affect the final yield much if the plants were allowed to grow until mature. For me, this was just a small insurance policy.

One day, the young teenager cut his forearm badly with my razor-sharp knife while slicing through a tough, fibrous branch. I heard him scream and ran over to discover a deep wound. The blade had severed veins and arteries and muscle, all the way to the bone. I wrapped the wound tightly and left him in Lee's care. Then I ran to Piedritas, some five or six miles away, with lungs burning, coughing years worth of accumulated gunk out of my lungs. I barely made it.

Someone from the town drove out in a pickup and picked the kid up. Then he sent someone else to bring a doctor from San Miguel to repair the wound as best he could in such a crude environment. I watched as the doctor worked. He began by sewing the muscle together a layer at a time, gradually pulling the tissue back together. It took about an hour to complete the job. Not once did he ask the young man how he'd been cut.

* * * * *

Late one evening, I saw something that sent us scurrying into action. A convoy of trucks—green trucks full of Mexican soldiers—drove by on the main road connecting Piedritas with the rest of Mexico. It passed within a half a mile of our marijuana patch. Cheryl and the kids just happened to be visiting that evening. Cheryl, Lee, and my children immediately left in the pickup, driving cross-country away from the road. They hid in a spot a mile or two away from the *jacal*. My oldest son, Dion, and I followed later, carrying what we could in our hands. We forced a young orphaned burro the kids called Sancho to pack a jug of water. It took several trips to bring all the things we needed.

We hid out in the desert for three days. Finally the soldiers left without coming out to the marijuana patch.

After their departure, Cheryl returned to Piedritas. Once again, I asked Oscar for permission to cut down the patch. Once again, he said no. The government minister I had previously met came out again and reassured me that everything was under control. I don't think even he was prepared for the heat that would come down when Glenda got through telling what she knew or when that information made it onto the radio waves emitted by XREY—waves I would pick up as far north as El Reno, Oklahoma, some years later from the bunk of a prison cell. I cringed as I heard the radio commentator ask, "What are gringos doing in Piedritas? And how is it that no one has a job in Piedritas but many drive new pickups?"

I continued to apply pressure on Oscar, telling him at one point that I was going to begin cutting down the patch whether he liked it or not. He gave in. The largest plants stood close to fourteen feet high by this time with trunks six inches in diameter and buds as large as a man's wrist. A large group of local men showed up to harvest, along with several trucks. One by one the plants were removed and stacked onto the beds of the trucks, which were equipped with tall side rails. By the end of the day, our patch was gone. My part of the job was done. I breathed a huge sigh of relief. Now all that remained was to wait for the marijuana to be cured, packaged and sold. Oscar and his men were going to handle all of this. Without anything to do at the *jacal,* Lee and I moved back to Piedritas.

Lee was still sad that Glenda had been caught and was worried about her fate. But he was glad that at least we would have enough money to move on once the marijuana we had produced was sold. I too hoped to move somewhere else with my family—the memory of that convoy of trucks full of soldiers was hard to erase from my thoughts.

* * * * *

The next morning Lee and I sat out near the pile of boulders on the edge of town, enjoying the morning sun and a joint, as was our custom, when Vicente, Oscar's youngest brother, came galloping up to us on horseback.

"Run! The soldiers are here!"

"What about my family?" I asked.

"They aren't interested in them. It's you they want. Run! I'll take care of your family. I'll take them to the shack where you and Lee stayed north of town."

In a state of shock and despair, Lee and I ran for the nearest mountain and climbed to the top without stopping. We selected a spot behind a large rock and watched from high above as the soldiers entered and terrorized the tiny town. We were witnesses to a living nightmare. We heard shots, women screaming.

People ran from the town on foot, clutching children. Engines roared. Vehicles raced around, raising clouds of dust. Chaos prevailed. All of this played before our eyes like some scene from a horror movie. We had wanted to help out the tiny community—instead we had introduced unspeakable evil upon them. Piedritas would never be the same.*

As night came on, we began walking north until we hit the first line shack where we had spent time guarding marijuana for Oscar while waiting for McCasland's never-to-be arrival. Vicente and my family were there, looking like frightened deer caught in the headlights of an oncoming vehicle. I told Cheryl to go back to the United States. She didn't want to go, but this time I wasn't asking. Vicente reloaded them into the back of a pickup and started north toward San Vicente crossing, using a back route that was rarely traveled. That night, it started raining—hard. Cheryl and my kids would have to cross the Rio Grande in full flood.

---

* The deal I made with Oscar would have given me 40 percent of the proceeds, with sixty percent allotted to him and the community.

ALONE IN THE DESERT

Celerino dropped Lee and me at some abandoned fluorite mines a few days after the soldiers left. The mountainous area was riddled with mineshafts. One of them contained about sixteen thousand pounds of Colombian marijuana. Lee and I slept on this huge stack of bales with our faces just a couple of feet from the ceiling of the cave. I was uneasy in the mine. No supports held the ceiling in place. Every noise spooked me and I couldn't help thinking about what would happen if an earthquake hit and sent the mountain down on us.

Oscar had been in Fort Stockton picking up some supplies when the soldiers arrived. He was in no hurry to show up in Piedritas after hearing what had happened. Eventually he returned and told us that Cheryl, Stephanie, and my children had made it safely back to the United States in spite of the fact that the river was in flood by the time they got there. Vicente described what had happened that day:

Oscar's men had moved our harvested marijuana plants to a location to be processed. Lee and I didn't know where it was. A small but well-trained group of soldiers arrived and knew exactly where to go, meaning someone had informed them where the men and the marijuana would be found.* The soldiers got the jump on Sergio and one of Oscar's brothers-in-law and ten other men from the town who were still in the process of cutting up, curing and packaging the marijuana. By chance, Vicente had gone to town on horseback to bring back supplies and was on his way back to the group when they arrived. He turned and fled on horseback to warn Lee and me, barely beating the soldiers back to town.

The soldiers treated the twelve captive men badly, at one point forcing them to run while they drove behind with loaded weapons aimed at their backs. They clubbed Oscar's brother-in-law in the back of the head with the butt of a rifle, knocking him unconscious when he refused to run any more. They yanked a good portion of Sergio's beard out of his face and busted both of his eardrums with simultaneous open-handed slaps to the side of his head.

They then forced the men to finish packaging a portion of the marijuana. They weren't delivered to the jail in Piedras Negras until three days their arrest. It took that long to finish their work. Mexican newspapers made a big deal out of the bust, complete with pictures showing piles of plants being burned. What failed to make the news was that most of the material being burned was the inferior portions of the plants.

No one was killed or detained inside the town itself, but all were terrorized, and some houses were looted. The soldiers had driven around firing automatic weapons into the air, helping themselves to whatever they wanted.

Oscar tried to pay to have the men released, but the bust had received too much attention. No amount of money would free them. They were transferred from the jail in Piedras Negras to the prison at Saltillo.

* * * * *

Our marijuana, the product of all of our labors, was gone, but still there was the Colombian. Lee and I helped Oscar move the Colombian marijuana from the mines to a natural cavern high in the mountains north and east of Piedritas. This left us once again with nothing to do but wait until David McCasland resumed flying loads, so we returned to the *jacal* where we had grown our marijuana. One evening we decided to make a trip to Piedritas to pick up supplies that were left in the house where Cheryl and my kids had lived. We walked the five miles to the

* Recently I learned that Macario Villarreal, Oscar's own cousin, was most probably the informant.

town with five-gallon buckets in hand, using back trails, and approached the town under cover of darkness.

Quietly and carefully, we snuck up to the edge of town to observe. We approached through a ditch of sorts, when the sudden cry of a woman startled us. She had two small children with her. Once she saw who we were, she breathed a sigh of relief.

"Son los pobres gringos. No se preocupe," she told the two small children at her side. ("It's just the poor gringos. Don't worry.") The woman was hiding in the ditch with her children, afraid to be in her own home. She was more afraid of the Mexican cops than anything in the surrounding desert night, including us—two escaped fugitive smugglers of illegal drugs.

Lee and I picked up what we needed from the house I had rented for Cheryl and the kids, loaded it into the plastic buckets and started walking toward the *jacal.* The buckets became very heavy over the distance and both of us struggled. My arms and shoulders ached. I fought to stay awake while I walked. Once again, I was surprised how tough Lee proved to be. I was pushed to the very limits of my strength and endurance to make it back, and Lee matched my efforts, stride for stride.

The odd thing about this situation is that the Colombian marijuana was never touched. Lee and I didn't know who to trust or what to think after that. All operations ceased for a time. Then Oscar sent word that McCasland was going to fly down and pick up a load. He and another pilot called Gil showed up, each in a separate plane. This fellow Gil was an intriguing sort—very much a professional dope pilot. I would have liked to ask him some questions, but knew there was no good way to do so.

McCasland arrived during daylight hours, but Gil landed late at night on the hard flat surface of the desert floor with nothing but a pickup illuminating the piece of ground he landed on. The tail-dragging Helo-courier he flew was capable of landing almost anywhere. With a stall speed of thirty-five miles an hour and an unusually stout frame, the plane was designed to survive a crash. Gil carried a 55-gallon barrel of gasoline in the back of the plane that he used to top off his tanks in the event he needed to turn around and fly back to Mexico. His plane was equipped with all kinds of radios, containing scanners that monitored all law enforcement frequencies. He knew the routes into the United States and how to avoid contact with radar and people. Each time he made a successful trip, he would stamp another marijuana leaf into his belt. A count would reveal a hundred leaves imprinted there and room for no more. He didn't drink alcohol and he didn't take drugs.

A whole group of others I did not know attended this event, among them several Colombians who colored all their sentences with the word *puta.* You'd think everyone and everything was either a whore or the son of a whore after listening to them talk. We smoked marijuana and took shots off a bottle of brandy. The weed we were smoking was some of what I had left from the patch we grew. One of them let it slip that he had purchased 800 pounds of the higher-quality bud from our patch. He said he sold it wholesale for $600 a pound, yielding $480,000.

Oh really! I thought. I looked at Oscar. He just shrugged his shoulders. I assumed he, too, had been fucked on the deal.

Lee and I helped load the planes and watched in envy as they flew off, destined for the United States.

Lee was anxious to use his flying skills and probably had seen all he wanted of the monotonous lifestyle we lived, so when McCasland returned a few days later, Lee flew off with him. I found it hard to watch him go. I knew I would now be the lone American left in Piedritas, but plans called for Lee to fly back and get another load. When Lee returned, I was going back with him to acquire supplies and a fake identity, which I hoped would enable me to move to and live in another part of Mexico. I just couldn't take staying in Piedritas anymore, with the constant threat of being captured or killed by Mexican soldiers.

I watched them leave, and then I waited. Day after day went by. Since Oscar had moved the remaining marijuana to another location, I was left with nothing to do, stuck out at the fluorite mine, waiting for Lee to return. But he didn't. I became bored and lonely. I wandered through the mountains, thinking and praying. I ate what I thought might be some peyote. It gave me no answers. Beto came by with supplies once a week. The rest of the time, I sat on a hill and watched, hour after hour, day after day.

One day, a patrol of soldiers walked by the entrance to the mine where I stayed. I was about a half mile above their location, hidden behind a rock. I remained in place, with my heart beating twice its normal rate and adrenaline coursing through my veins. I was reminded of playing hide-and-seek as a kid, only now my life was at stake. They missed me. From that day forward, I slept in a different place each night.

I had a sleeping bag Cheryl had brought down, a mummy bag, which zips up around your body. When fully encased in the bag only a small portion of my face remained exposed. I stared at the stars until I fell asleep, hoping some rattler would not be attracted to the heat my body put out. After an hour or two, I woke up and listened—without moving—for any unnatural sounds. Hearing nothing, I fell asleep again for another hour or two, and then woke up again to repeat the process.

Lee had taught me to identify a number of constellations. I knew what stage the moon was in on any given night and could tell what time it was by simply looking at the sky to see how far things had rotated. I closed my eyes. An hour later I woke up again, looked at the stars and listened. I couldn't relax anymore.

Beto didn't return for a number of days after the reappearance of the soldiers. I began to worry that he too had been arrested or that citizens of the town who had lost loved ones might turn me in, hoping to secure the release of their own. I ran out of food and coffee. Lee had once told me he got headaches when deprived of caffeine, but I had never experienced that until then. I got a hell of a headache, which may have been compounded by the fact that I also went several days without eating anything. I finally got tired of sitting there, doing nothing, and decided to take a hike toward Piedritas. I spied on the town from above. Things seemed normal, but I didn't want to enter the town, so I bypassed it and walked to the irrigated fields below the reservoir. I was extremely hungry by this time. I found some sorghum-like cane which had sugar-filled sap in the stems. I chewed the stems, spit out the pulp and swallowed the juice for energy. I walked back to the *jacal* where we had grown the marijuana to see if anything had been left behind. I found a coffee pot and a small amount of coffee. I also found about a pound of white rice and a pot. That evening I had a meal. The coffee instantly took away my headache.

I decided to stay at the *jacal*. Day after boring day passed until finally one day Beto came riding up. Lord, I was glad to see another friendly human face and to hear his voice. I had gotten to the point where I was talking to myself. He brought me a few supplies and then left. The loneliness and isolation continued. There was no word about Lee. I began to be afraid he was dead.

I prayed and got no answers. I shouted at the sky around me and got no answers. At one point, a very unusual thing happened. A coyote approached my camp and walked up to within twenty feet of me. In all the time I had been there, if I saw coyotes, they were running to distance themselves from us. But this one approached me with no fear whatsoever. His demeanor wasn't threatening. He just stared, and I stared back. Somehow he sensed I wouldn't hurt him, and I knew exactly how he felt. We shared a common bond—we both were despised and hunted. He most probably had watched me for days. I tried to run him off, but he retreated only a short way, staying just out of reach until I stopped and then he stopped—and turned and stared again. When he did decide to leave, at least ten minutes later, he simply turned and walked off. In a way, I guess God did finally answer my prayer. But I wasn't sure of the reply other than the feeling I was left with, that maybe He did know where I was and how I suffered. I took comfort from that. I also decided I was going back to the United States, one way or another.

I sent word through the next available contact to David McCasland, telling him I would be going home the next time he did a trip.

PABLO ACOSTA AND THE DEATH THAT WASN'T

In the meantime, Oscar decided I could be of some use, guarding what was left of the DC-6 load of Colombian marijuana. I was glad to have something—anything—to do. Beto took me to a new camp, high up in an arroyo. I placed my things in a small cave and settled in. He left me with supplies. And then, once again, I found myself alone in the desert mountains, with no one to talk to and nothing to do.

I scooped a small pot of water out of the *tinaja* near my cave, carefully avoiding the creatures swimming around in it and set the pot above the small fire. A thin stream of smoke rose into the air from the *quebrada* that was my home. The *quebrada* was a steep-walled gorge that separated two ridges of the same desert mountain. I slept in a small cave about forty or fifty feet above my location in the bottom of the ravine.

My cave was really not much of a cave, but instead a slash in the face of the rock wall, five feet high at the opening, tapering as it went inward, twelve feet wide and fifteen or so feet deep. In it, I had found signs of Indians long since dead. Getting in required crawling. Opposite my cave—on the other side of the *quebrada* and at about the same elevation—ran a dim trail, leading up and over the ridge. If you were to follow this trail for a quarter of a mile or so into the next *quebrada* and knew where to look, you would find a small hole in the mountain, five feet in diameter.

A makeshift ladder descended from this opening to a landing thirty feet below. If you climbed down this ladder, you would find the opening to a series of large caverns. The air in these rooms was hot—in the 90-degree range—even though the outside temperature was cool, as are most December days in the northern mountains of Coahuila. A short walk into one of these rooms with a flashlight revealed a stack of rectangular-shaped bales of Colombian red and gold marijuana, each covered with multiple layers of packaging to protect the valuable contents.

I knew all of these things, of course. That's why I was here after all—to guard this weed. Or so I was told.

But who was I guarding it from? I often wondered. I had been here for over two weeks and the only person I had seen was Beto. Oscar sent Beto from the tiny village of Piedritas to bring me supplies. The supplies were meager, but appreciated—things like flour, lard, pinto beans and baking powder for tortillas. He usually brought eggs and occasionally tomatoes, peppers and onions. Always, there was instant coffee—thank God—and usually some sugar.

One day, Beto came by on horseback to check on me early, telling me he would return that evening with a load of supplies. I watched the beans boil like I did most every day. When they needed water, I added it to them from the coffee pot, just like he had taught me. I didn't tire of eating the same thing, but I was tired of being alone—and when I say alone, I mean alone. Since the soldiers had arrived to bust my marijuana patch, I had been on the run. I spent days and weeks without a soul to talk to. I was starved for human companionship.

I rolled another joint, fired it up and drank more coffee. Finally, I got the urge to take a leak. I got up slowly from my position in front of the fire, stretched to get feeling back into my legs and walked twenty or thirty feet down the *quebrada* on the downhill side of the *tinaja*—the only source of water I knew of for miles—unzipped my fly and pissed. I zipped my pants. Then it happened—*BANG! BANG! BANG!*

Three shots passed right over my head from close range, so close that I felt the wave of pressure as the bullets zipped by. I ducked. A rush of adrenaline coursed through my body as I ran for the cover of a big boulder to one side of the *tinaja* and dove under it.

BANG! BANG! BANG! More shots. I smelled burnt gunpowder. The shots came from the side opposite my cave, probably from the trail. The boulder under which I cringed was large, over ten feet tall. Water had washed out an opening in the sand and gravel. I wedged myself into that opening. I knew they couldn't hit me from where they were, but I worried what would happen once they figured out how to work around to the other side—the side where my cave was located. From there, I would be a sitting duck.

Who were they? How many were there? Why the hell are they shooting at me? I spotted the long-barreled .22 revolver Beto had left me for hunting. It was about four feet away, out in the open. I took a chance, darted out, grabbed it and dove back into my spot. My action was greeted with more gunfire. Then the men tried to bounce bullets off the rocks opposite my position, attempting to hit me with a ricochet. I felt rock chips or bullet fragments, but none hit hard enough to break the skin through my clothing.

Below the trail from which they fired, the *quebrada* was too steep to descend, but I knew that farther up the ravine, a couple of hundred yards, there was a place they would be able to get down. I was pinned. I couldn't move.

I decided to let them know I was armed, even though I had nothing but a .22 revolver. I wanted them to know they risked being shot if they moved into a position where they could hit me. I took aim at a rock and fired, trying to hit the rock in a glancing manner that would cause the bullet to ricochet, as I had often practiced, just for fun. The report of the .22 sounded puny compared to the loud explosions of their weapons, but the ricocheting bullet did what it was designed to do. My shot was answered with another barrage of bullets and a startled stream of cussing. I heard two people arguing, the first proof other than gunshots that there were people attached to the rifles. From their frantic speech I figured out that they were very unhappy they had missed.

I didn't want to shoot anyone, but there was no need for them to know that. I let loose with another round and heard more cussing. A rock rolled from above. Now I wasn't the only one scared. A few minutes passed. I surveyed my position. Sooner or later they were going to figure out how to get across to the other side. I had to do something.

I yelled out in Spanish, "Who are you? What do you want?"

I heard mumbling but no reply. I decided to take another chance.

"Are you here for the marijuana? If that's what you want, you can have it. All I want is my life."

I heard more talking.

I continued my one-way conversation. "Are you police?"

Finally I got a reply. "Yes, we are police!"

I decided to take a huge risk.

"I'm going to throw my gun out if you give me your word that you won't shoot me. You can have the marijuana. All I want is my life."

"Come on out with the gun in your hand," came the reply.

"What?" I asked, shocked at the order they gave. "I would rather come out unarmed!"

"No! Come out with the gun in your hand!"

This didn't make sense to me. No cop in his right mind would issue such an order. *Did they have a problem shooting an unarmed man?*

"Are you sure?" I asked, incredulously.

"Yes."

"Are you going to shoot me?"

"No. Come on out!"

I rose to my feet, holding the grip of the revolver between the thumb and index finger of my left hand, stretched both hands high into the air and walked into the opening, fearing I would be shot. I got the first look at my assailants.

There were two perched on the trail above me. Both looked young, way too young to be cops in my country and they wore street clothes. Both had weapons trained at me, one a semiautomatic rifle of some sort and the other a smaller automatic machine pistol. I placed the revolver on the ground.

"Pick it up!"

"What?"

"Pick it up!"

"I don't want to," I yelled in reply.

"I said pick it up!"

I picked up the revolver again with my left hand, holding it between my thumb and forefinger in such a way that they could clearly see I couldn't fire it.

While the man with the rifle guarded me, the other young man worked his way along the face of the ravine to a place where he could get down and walk back to my location. I had to give him directions. He approached and took the revolver. Then he guarded me while his companion made his way up the ravine and returned to our location.

"Where is the marijuana?"

"Up there."

I pointed at my cave. There was two hundred pounds in the cave, but it was lower-quality, early-cutting stuff, taken before reaching maturity from the ill-fated patch Lee Cross and I had raised. At gunpoint, I led them to my cave. They looked through my meager belongings. The weed was stuffed into the farthest recess of the small cave. They dragged out a couple of sacks, opened them and looked at it. Alongside my sleeping bag were a couple of pictures of my children. One of the men held one up.

"Who is this?" he asked.

"Those are my children."

"You have kids?"

"Yes, six and a seventh on the way."

The young man with the rifle kept it trained squarely at my face.

"Take it easy, man," I told him. "I don't want to hurt anyone."

"Why did you shoot then?"

"Just to keep you from coming around the rock and shooting me. I wasn't even aiming in your direction!"

I restated my position. "Take the marijuana. It will do me no good dead. All I want is my life. Please, don't kill me."

I had hoped when such a thing happened to me—if something like this happened to me—I would be courageous. I was scared shitless.

"It is not up to us."

They talked among themselves for a while. Finally, they began to relax. The one with the Uzi-like weapon left to go see the *comandante,* while the other stayed behind to guard me. The first young man was gone for about thirty minutes. I talked with the second man during that time. I told him how I came to be at the cave. At one point he asked about the *other* marijuana. This was the first indication they knew about the second cave.

"How much is there?" he asked

"About a ton, I guess." There had been more, but most was already sold. In fact, if they had arrived a day earlier, they would have found another ton. At that very moment, Oscar was at the river packing it into life rafts headed to the United States—to plenty of anxious smokers.

The second assailant returned with orders to take me back to see the *comandante.* I was led at gunpoint up the trail toward the large cavern. A hundred yards before reaching the crest of the ridge, I was blindfolded with a bandana and my hands were tied behind my back. The remainder of the walk was difficult, but I could see directly downward and managed to make it without falling. When we got to the top of the ridge, I heard more voices. I heard the approach of another man and made a big mistake. I tilted back my head and peered through the cracks near my nose into the acne-scarred face of a man—a man I believed to be Pablo Acosta.

"Don't look at me!" he shouted.

The two young men who had originally captured me quickly pulled the blindfold farther down.

The questioning began. "You lie, you die! Understand?" the leader began.

"Yes."

I stood facing him, with my original captors on either side. Behind my precarious position was a steep drop-off. I knew weapons were trained at me. I felt them.

"Are you alone?"

"Yes."

"Why are you here?"

"To guard the marijuana."

"For who?"

"My friends."

"Who are your friends?"

"The people of Piedritas."

"Which people?"

"All of them."

"All of them?"

"Yes. They all know what goes on around here." In reality, I suppose this was true, but I was trying to avoid giving him specific names just in case they really were cops.

"Don't bullshit me! I will kill you! Now, who owns this marijuana?"

"Some Colombians. They made a deal with the people of Piedritas."

"Who in Piedritas is in charge?"

I could no longer avoid telling him what he wanted to know without jeopardizing my life. I wanted to live.

"Oscar."

"Oscar who?"

"Oscar Cabello."

"Where is Oscar?"

"He's down at the river, loading some people."

"So you are alone?"

"Yes."

"Do you expect anyone?"

"Yes."

"Who?"

"Beto."

"Who is Beto?"

"He works for Oscar. He's going to bring food this afternoon."

"Will he come alone?"

"Probably."

I heard the man issue orders to several other men in the distance. Then I heard them discussing something among themselves. The *comandante* returned. He ordered me moved to sit on a pile of spent *candelilla* plants. A guard was left to watch me and the rest of the men nearby sent to set a trap for Beto. He told the man guarding me, "Shoot him if you hear a shot."

"Wait a minute!" I objected. "They shot at me totally unprovoked!"

"Shoot him if you hear a shot!" he repeated, and then added to me, "You shut up!"

"Sí, mi comandante," my guard replied. I shut up.

Soon we were alone. I waited and prayed, waiting for the sound that would signal my impending death. I went over all the things that had brought me to this place. I couldn't get over the fact that I was about to die over weed. I prayed for forgiveness of my sins. I prayed for my family and I prayed for courage. I prayed for my captors. Finally I began to talk again with my guard. A couple of hours went by and there was still no shot. The *comandante* returned. He asked more questions. Finally, I heard the approach of a couple of men.

"We got him," they told the *comandante.* Apparently they had captured Beto without incident. I breathed a sigh of relief. The relief didn't last long.

I overheard the *comandante* talking to some other men about the marijuana and about me. He couldn't decide what to do with me.

"How much marijuana is in the cave?" he asked me.

"A ton," I replied.

They were having a difficult time getting all the bales out of the cave and loaded into their vehicles. Putting the bales into the cave originally had been an arduous task, involving many men positioned on the ladder in a human chain. There was barely enough room to stand on the ladder and pass the bales by your body through the narrow vertical shaft. The *comandante* sent his men out again and I was left alone—with him. At some point I managed to rework the blindfold up a little by raising and lowering my nose. I leaned my head back again and caught another glimpse of my captor. He was standing about fifty feet from me with an AR-15 rifle propped over his knee. He kept it pointed in my direction. Luckily, he was smoking a cigarette and looking off in another direction when I stole the look.

From this distance, he began to question me as though he was trying to decide whether I should live or die. He wanted to know how a gringo ended up in that part of Mexico.

I told him about how I had escaped from prison, the failed attempt to grow and sell marijuana, which ended up in a bust, and the months I spent hiding in the mountains from both governments.

"Have you ever killed anyone?" he asked.

"No."

"Never?"

"No."

"If I let you live, will you come after me?"

"No, sir. All I want is my life."

Finally he made his decision.

"I am going to let you live. I am going to leave the marijuana you grew in the cave to give you a new start. You take that back to the United States. You have no business down here. Go back to your country! Do you understand?"

"Yes, sir. I don't even care about the marijuana."

"Take it as a gift from me."

"Yes, sir."

A few minutes later, my original captors returned and led me back down to the *tinaja,* removing my blindfold shortly after the walk began. When we arrived at the campsite, I saw Beto sitting on the ground with his hands tied behind his back and a very worried look on his face. I said nothing.

No matter what the *comandante* said, I still wondered, *Were we to die?*

I was ordered to sit beside him. The men walked away.

After about fifteen minutes of silence, Beto and I began to talk. I wanted to know who had done this. He had seen no one other than some of the lower-level people. While he wouldn't tell me, I suspect he knew who they worked for.

I remain convinced this was Pablo Acosta and his men and that their actions eventually contributed to his death. Pablo may not have known it, but he and Oscar worked for the same folks. Oscar never recovered the marijuana taken that day. The vehicles left the area headed toward Santa Elena, Chihuahua. Few men were capable of selling such a quantity of merchandise as easily distinguishable as Colombian marijuana without it coming to the attention of everyone in that region.*

I don't believe my captors were real policeman because of the way they handled my capture. No cop in his right mind would order me to pick up a loaded weapon, especially when I had offered to throw it down.†

As for the marijuana, it really meant nothing to me compared with the possibility of losing my life in the northern mountains of Mexico. But he insisted I take it. Air interdiction officers of the U.S. Customs Agency near Tucumcari, New Mexico, arrested me with this gift a short time later. That proved to be worth an additional eight-year sentence on top of the seven I had left behind when I escaped from prison.

What a gift!

Thinking back to this incident, part of me died on the side of that mountain that day.

---

* Colombian red and gold marijuana is radically different from anything produced in Mexico. It has a unique odor and look. It's packaged differently and easily distinguishable, even to a novice user or dealer.

† I know the man questioning me was debating whether or not to allow me to live. It could have gone either way. He chose to show me mercy. I don't know why, but I thank him for my life wherever he may be. If it was Pablo Acosta, he's not alive to receive my thanks. Pablo died within a year of this incident in a gun battle with Mexican authorities in Santa Elena. The Mexicans had been dropped into the town by helicopters coming from the American side. Through Oscar I learned that Amado Carrillo was at Santa Elena on the day Acosta was killed, flying above the scene in a plane. He paid one million dollars to see Acosta killed. Some credit Pablo Acosta and his organization with over fifty murders. I consider it a miracle that I am not counted among his victims.

BACK TO JAIL

There are certain days in a person's life that prove pivotal. December 17, 1986, was one of those days for me. I started the day in the *despoblado*—the badlands—of northern Mexico and ended it on a hard metal bunk of a jail cell in Los Lunas, New Mexico, surrounded by the noise and confusion of too many people crammed into too small a space.

I had been completely alone for months in the mountains of northern Mexico, except for occasional contact with Beto. I sent word to David McCasland that I wanted to come home. I had two hundred pounds of fairly low-grade marijuana, a "gift" to me from Pablo Acosta and his *bandidos* after they relieved Oscar Cabello of about a ton of Colombian marijuana. I almost found it surprising that Oscar was going to allow me to leave the country, knowing what I knew. The fact that he did spoke well of him and his dad.

David McCasland arrived at a landing area near Piedritas in another Cessna 206 turbo.*

---

* I learned from David that Lee had started back toward Mexico to retrieve me and another load several days after their last trip, but had disappeared. David read that Lee had been forced to land at the Midland-Odessa International Airport when an electrical fire filled the cabin with smoke. It was discovered he had no license. All the seats had been removed from the plane and a few seeds were found. He still wore the watch he owned while in prison, a watch with his prison identification number engraved into its back.

After all I had endured in Mexico, I was glad to see his face. David had never smuggled marijuana before he met me in Big Springs, but he did know how to fly a plane. He hadn't learned much about his new trade. He arrived without any extra gas and hadn't made arrangements with Oscar for any either. The Cessna did have tip tanks and a Robertson STOL kit, so he managed to get it on the ground. We had enough fuel to make it back to Tucumcari, assuming all went well. I'd seen all of Mexico I wanted. I loaded the weed into the back of the plane and we took off.

David began following the road out of the park, which was crazy. We flew directly over a law-enforcement vehicle headed north, almost as if it was waiting on us. Maybe it was.

The weather was clear when we left Mexico, but soon after we cleared the park we ran head-on into a blizzard, blowing straight out of Canada. Afraid we might hit something like a radio tower or a mountain, McCasland climbed to ten thousand feet and flew on instruments. He hadn't been flying nearly low enough to suit my fancy before this, but once all visibility was gone even I saw little choice in the matter. I sure as hell didn't want to crash.

When we got near our destination, McCasland began to descend. I hoped he knew where we were and that we wouldn't hit something. Finally we broke out of the clouds, very near the ground. When we did, another plane appeared, flying in the opposite direction. We crossed paths, much like two cars headed down opposite sides of the interstate. This freaked me out, but didn't seem to bother David much.

We continued on until we reached a field on a farm near his house. We landed in the snow-covered field and unloaded the marijuana. As soon as we hit the ground, I heard something come over the radio with the word marijuana in it. I knew we were screwed.

David appeared not to have heard this. I shouted at him and told him to take off again. He did. While we flew, I looked at the fuel gauge and calculated. We could not make it back to Mexico. I wanted to land somewhere remote to see if we could escape on foot. When we descended from the clouds again, I could see law-enforcement vehicles scrambling to cover the ground below. I didn't know it at the time, but law-enforcement officials had installed a transponder in the plane to track us. The paperwork on my case would later reveal McCasland had bought the plane for sixty thousand dollars, all cash and small bills, some as small as fives. Why anyone would think that suspicious is beyond me!

Up again we went, higher and higher, until finally we broke out of the winter clouds. When we did, another plane appeared on my side of the plane. The pilot made hand signals to us and held up his radio microphone, trying to tell us what frequency to go to. I ignored the guy.

We flew merrily along like this for a few minutes, like two cars cruising down the highway, side by side. Then they got my attention. Our plane lurched upward, feeling like a launching rocket. A second later the tail of a King Air appeared in front of us, shooting skyward, right in front of our prop. When we hit the dead air left in the wake of this King Air, we fell. The fall almost ripped the wings off of our plane. The crazy bastards had made a run at us from below and behind at full speed, and then pulled up right in front of us.

A few minutes later, they did it again. After the third or fourth time, I told David to get on the radio and tell them we would be going to the airport at Tucumcari to land and give ourselves up. And that's exactly what we did. We landed at the airport and taxied to a stop. A helicopter circled overhead and the chase planes landed behind to join a shitload of ground-based cops ready for our arrival. Even the airport security guard got in on the action with his shotgun.

David and I remained seated in the plane. After a minute or so, a group of heavily armed cops, with weapons drawn, approached the plane, ordering us to get out. The plane's engine was still running. David turned it off and got out, but there was no door on the passenger side and every time I tried to lower my hands to release the seatbelt, a cop standing on my side of the plane started tightening up on the trigger of the shotgun he had pointed at my head.

He shouted again, and again, "Get out!"

I wanted to get out, but couldn't do it without releasing the seatbelt. We went through this process several times before I was able to shout loud enough to make him understand my predicament. When I finally did get the belt unbuckled, I climbed over the pilot seat and out the pilot side door. My first taste of American soil in almost a year was face down into the frozen tarmac of the runway with a handful of cops on my back. I had no coat and it was cold as hell—the wind was blowing thirty miles an hour. Eventually they got us up and took us into the airport.

After an hour, we were loaded into the larger twin-engine Cessna that had originally pulled alongside us. We rode with hands cuffed, seated facing two of our captors. The blizzard still howled in full force. All of us were somewhat apprehensive—at least I know I was. I overheard the conversations of our captors on the way. Most of these guys were Viet Nam veterans who had gotten hooked on the excitement of dangerous flying during the war. Their skill and courage was phenomenal. I remember a look from a couple of them—almost expressing gratitude that we did what we did so they could do what they did. There was no hatred—or at least that was my take on the subject. I got the feeling they would have been tempted to try the other side of the fence had not the legal option to participate been available to them.

One of them asked me what I thought about their tactics.

"Scared the shit out of me," I told him.

"You didn't see nothin' man. If you'd a tried to make it back to Mexico, we'd a really shown you a thing or two."*

That night as I lay down to sleep, I was relieved. I had survived against all odds. I also had gained a healthy measure of respect for members of the U.S. Customs Air Interdiction Teams.

---

* Later, while in the county jail at Las Lunas, another group of would-be smugglers arrived with stories much wilder than ours. Apparently they did have enough gas to make it back to Mexico and tried. After being buzzed several times from below and behind, as we had, that big King A air-descended upon them from above, almost landing on their little Piper in mid-air, robbing the lift of the little plane and driving them all the way to the ground where they crash-landed.

FREE AT LAST

DEA agents transported David McCasland and me from the airport in Albuquerque to the county jail in Las Lunas, New Mexico. We were stripped and searched and given orange jumpsuits. Then they separated us. He went to a cell on one side of the jail, and I was sent to another at the opposite end of the jail. There was one available iron bunk in the cell where I was deposited so I placed my vinyl-covered mattress on it and lay down.

Outside, it was cold. Inside, it was warm. I looked at the concrete ceiling just inches above my face and tried to sleep, but my mind would not allow it. I felt safe for the first time in months. Being around other people was a strange sensation, a good sensation. Most were unhappy to be in that jail. I felt relief. I had reached a place where I could now quit. I listened to sounds coming in over a radio and heard new songs. I eavesdropped on the conversations of others.

Then, from a nearby isolation cell, an eerie cry began, the voice of a man considered crazy by my cellmates.

"Not again!" one of them shouted. "Shut up, motherfucker!"

The man continued in a song-like voice, as though he were describing a living dream. He described bombs and fire and screams. He yelled and begged and cried and screamed—from deep within. The man was in Viet Nam. And so was I.

His cries were interrupted only by occasional shouts and threats from my cellmates trying to get him to shut up.

After several hours, guards tired of hearing this went into his cell and hog-tied the guy, using cuffs on his hands and running the chain from a set of shackles from one foot, through the handcuffs, and then to the other foot, behind his back. He remained hog-tied in that position the rest of the night but continued to cry and scream and describe his living nightmare. I shared in his agony.

The following morning I was moved to another cell containing fourteen men. Breakfast was served to the bitter complaints of everyone except me. I thought it was great. I had taken a huge step up in the world.

I applied for and received counsel from the federal public defender's office. I hadn't seen or heard from my attorney, yet I sat in court awaiting arraignment. *Here comes another fucking,* I thought to myself. My previous experience with Steve Rogers had left me distrustful of all lawyers. At the last moment, my attorney appeared—a woman by the name of Ann Steinmetz. She took a seat beside me and to my surprise vigorously defended my defenseless position. *What is this, some kind of joke?*

Ann Steinmetz visited me at the jail a couple of times and did her best to defend my position. Never once did she solicit money from me. She restored a measure of hope in me that there are people working in our government who are truly interested in justice. She not only defended me publicly but took the time to get me out of the jail for a little conversation I very much appreciated. Thank you, Ann. You are not forgotten.

I received eight more years to be run consecutive to the seven I had left behind when I escaped.

* * * * *

A joint went around the cell. When it was offered to me, I declined. I still decline to this day, some sixteen-and-a-half years later. And I don't miss it, not one little bit.

AFTERWORD

I walk out of Davila's Barbecue in Seguin where I go to eat real food—the food of the poor—ribs, brisket and sausage—all the poor-quality leftover pieces the rich don't want. Davila's has no prime marbled cuts of loin or ground round beef, yet there is a richness and body to the food found lacking in the restaurant high on top of a glass building not so far away, where all is silver and glass and fine linen and painted women and soft men in their loafers. At Davila's is found smoke and dirt and boots and wood and fire—oh yes, fire—that magical stuff without which none of us would be: here I find life.

I put the key into the ignition and turn it.

My damned truck won't start.

Heat has paralyzed the engine. I reach for my cell phone, but it's not there. I left it on the desk, two miles away, connected to the charger. I have work I need to do. Right fucking now! Damn it! I walk into the restaurant again and tell Mr. Davila I have to leave my truck until it cools down. I know it will start after it cools, but not before.

He smiles—he's one not so far removed from the real world. He lives with the fire and the smoke and the roasting flesh.

"Can I give you a jump?" he asks.

"No, it just needs to cool down."

"How about a ride?"

Here is a man with work of his own, but he remembers what it is like to need.

"No, I'm okay. I'll be back later."

I walk out the door, look at my watch and begin walking. It's two miles to the office on the horse farm where I live and work. I turn right where I always turn right, but now I move slowly. Each step means one yard, one second of precious time. I look again at my watch. I walk. I look at my watch. And then an old familiar feeling returns, step-by-step, breath-by-breath. I feel the heat of the sun—the sun I so diligently hide from. My body warms with the heat of exertion and for a minute I am afraid. I haven't done this for some time. Will I faint or pass out from heat stroke? And then the sweat begins to form. The breeze tickles me and I am alive. I walk past a ditch I fly by in my iron beast every day, and I notice the weeds and the flowers and the trash someone so haphazardly has discarded. I cross the railroad track and smell the melting asphalt. Cars zip by, looking but not looking, speeding toward who knows what? But always speeding.

I pass a field and see the wheat. How is it that I failed to see this? I have passed by here many times. I break a head from the wheat and see the cracks in the earth. The grains are small. We haven't had enough rain this year. I crush the head in my hand and eat a few grains. It's good. A combine moves through the field, but the man operating it isn't thinking of the grain, the life-giving grain. He thinks of money or rather the lack of it. I see another field. Corn. I've seen this corn in passing and noticed it hasn't grown as tall as it should be by now. But now I see it clearer. The earth is dry and cracked. The leaves fold in upon themselves, trying to hide from the relentless rays of sun. Along the edges they are brown and scarred. The field is crying. How is it that I haven't noticed this?

Now the sweat soaks my shirt, but the breeze feels good blowing through my hair. I near the interstate. I have to cross this to get home. I watch as the cars fly by. The people see, but they don't see. They rush madly, insanely, oblivious to the world around them. I know this highway connects cities, from coast to coast. Along its breadth, people streak by—seeing—but not seeing. The huge trucks speed by as well—the blast of air and smoke and filth hits me in the face with their passing. I wait for a break.

I see my chance and run.

I make it.

Now I'm back at my computer. Thirty minutes have passed, and I remember what I should say. I look at the shit I've written, and it is shit. We have forgotten where we came from.

I must remind myself.

I must not forget.

We must not forget.

But we do.

For some reason we look for the answers in drugs.

* * * * *

I drive by the field of corn. Since the day I was forced to walk, forced to see, I know the plants. Rain came but too late. With it, the corn—which had fought to survive—grew, but the drought took its toll and some plants failed, leaving spaces in the rows. Those that did survive now spend their last energy producing the seed which will ensure another generation of the species.

A tractor pulls into the field, behind it a shredder. A man sits in the air-conditioned cab. The tractor starts through the field and slaughters the plants by the thousands, leaving a mat of tattered remains behind. I see broken cobs with yellow rows of corn. Somewhere people starve, yet the farmer has to plow this food into the ground for economic reasons.

The talk around the feed store is of fuel—how we may be able to make more money if we turn the corn into alcohol to power our cars and speedboats, so there is hope. Farmers may make money growing corn after all.

* * * * *

A mare lies on the ground, pushing, sweating. I wait behind her to catch the foal. I see the front feet and feel the nose. She pushes again and I begin to pull, fighting to maintain my grip on the slippery front legs. I press against her buttocks with my feet and arch against the load. The foal begins to move, ever so slightly. She stretches.

I've seen this hundreds of times, yet each time I'm amazed. The foal looks at me even before clearing the canal. He fights for a breath of air. The mare turns and looks at me. She trusts me. She won't trust another horse, but she trusts me. But she is old. She has given me many foals before this. Now money rears its ugly head. Do I continue to feed her at one thousand dollars a year? Everyone tells me she should go to slaughter.

* * * * *

Another mare I love gets sick with colic. I give her a shot of Banamine in the jugular vein, but she doesn't respond. I give her more—still no response. She looks at me, almost begging for help. The last mare I took in for colic surgery cost me four thousand dollars, four thousand dollars I don't have.

I watch as she dies a torturous pain-filled death.

A few days later, the scene repeats itself. I can't stand to watch another mare suffer this fate. I look for a way to end her suffering but, because I am a convicted felon, I have no gun. I have no drugs, which would end her suffering mercifully. She thrashes on the ground and groans. Her hooves cut her legs and blood splatters the wall of the stall. I get a plastic garbage sack from the kitchen.

My wife, Leah, asks what I am doing. I tell her she doesn't want to know and go back out to the barn and the mare's stall.

I rub her neck and talk soothingly and then slide the sack over her head and pull the drawstring. I watch the plastic stretch into her nostrils as she fights for air, but I hold tight and wait as she struggles violently—murdering her. After she dies, I fight back tears. But they come anyway.

* * * * *

Aguilar leads my filly into the paddock. She looks out at the crowd. I see her legs twitch and shake from nerves. I place the chamois on her back. The valet accommodates the jockey saddle. The cinch is tightened, the over-girth stretched into place. One last look to be sure all is in order.

I walk to the paddock. I tell the jockey how I want him to ride the horse. He nods, but both he and I know that once the gates open all of this will be forgotten, and he will just ride like hell. I leg him up and walk to the grandstand to await the race.

Leah and I take a seat in the bleachers. I survey the crowd. Blacks, whites, browns and yellows—tall, short, old and young, scantily dressed sexy young girls who spent hours grooming themselves. Two lesbians smoking cigars. A couple takes a seat in front of us. Four children accompany them. One, a brown-skinned young boy has eyes that indicate he is something special. I wonder—will he be the next Amado or the next Fox? A young girl, maybe three years old, stands next to her mom, watching the horses go by. The man and his wife appear to still be in love. The girl looks up at her mom in adoration and her mom hugs her. Then the girl turns and looks at me and smiles—an innocent, beautiful smile. I smile back, looking into two big brown eyes. And then I see it—she has a cleft palate. Behind my smile, I almost want to cry. She doesn't know she has it, but we will be sure to let her know as time goes by.

Suddenly who wins the race seems unimportant.

And I realize I only have two choices in this world—to love or to hate.

And the love of money and the things it can buy is behind most of what's evil in this world.

SEPTEMBER 2003

Just recently, September 2003 in fact, I had the opportunity to visit Piedritas with Charles Bowden, author of *Down by the River*, the best book ever written on the subject of Mexican drug smuggling, and Julian Cardona, a Mexican journalist and a photographer. They had both read this story.

I was nervous about the trip for various reasons. I had not left the United States since being released from prison and many years had passed. People from Piedritas had gone to prison because of my activities. I feared a new group might be working out of the area and that they might think we were narcs, looking for information. My fears proved unfounded.

Oscar and his family received me as one of their own.

Here in the United States, any reference to my past activities is quickly followed by a "but I don't do that anymore" and a variety of other disclaimers designed for acceptance. Such disclaimers were unnecessary in Piedritas. In fact, I was received almost like a returning war hero. I hardly knew how to react. Any attempt at an apology was quickly brushed aside and I felt no ill will or hard feelings from those whose lives had been so adversely affected by my activities.

The *ejido* had shrunk from around seven hundred people to sixty. Most of the young people had fled, some to *maquiladoras* near the border, others to work here in the United States. Among those who remained were some very dear faces.

Beto is now sixty-three years old but still spry, as we would discover while trying to keep up with him in an excursion through the mountains. His skin is weathered, his hat slightly askew like I remembered. His eyes still smile. He makes his living herding a few cows.

Others still harvest *candelilla* and sell the wax, which is now worth slightly more than when I lived there. Still others—whose names I had forgotten—appeared, and we reminisced about the old days. Among them were Reynol, Kiko and Eliezer. The smiles on their faces and the love in their eyes was unmistakable.

I found myself wanting to help them again, offering advice on how they might be able to better use their fields. This is a trait most Americans share—we have all the answers, if the rest of the world will just listen.

But that night, after the solar-powered fluorescent lights now illuminating the homes had been extinguished, I realized *they* have lessons *we* need to learn. I felt a part of the sense of community where everyone shared—not only their meager possessions but also their time and their love for each other. This in stark contrast to many American communities where people scarcely know their next-door neighbor. I slipped out and surveyed the beautiful night sky, a sky that no longer exists near the polluted place I now call home. I remembered Lee Cross showing me the various constellations. A peace enveloped the community. I was acutely aware of the presence of my neighbors.

We had arrived prepared to camp out, but instead we were given a house to stay in even though we arrived late and unannounced. Early the next morning, before first light, roosters crowed and a burro announced his presence with a long extended bray. A dog barked. The rest responded in unison, each prepared to vigorously defend his or her piece of ground.

Later that morning Beto gladly guided us around the *ejido* and invited us into his home for meals. We revisited the site of my capture. I saw a *candelilla* bed in the cave and then was shocked to discover that the bed was mine, looking like it had been left only the week before. No one had used it for seventeen years. Flashes of that day rushed to the forefront of my mind—the interrogation, the sound of the shots, the smell of gunpowder, walking blindfolded and the prayers I said while I waited to die. I could almost see the scared faces of my captors. I noticed the spot where the pictures of my children had been and remembered describing them to my assailants. Beto had not returned to the site since that day either and I could tell he too shared some of those flashbacks. As he pointed to the place where he had laid, tied and blindfolded, a trace of the fear he felt that day flashed across his face. People aren't supposed to survive such an event. We both knew that. But we did.

We went to the site of the well where Lee and I grew the marijuana. The *jacal* had fallen down and the well looked different. The men had refurbished it and it seemed deeper. Now it serves as a source for drinking water for cattle and also for a nearby *candelilla paila.* A new *jacal,* made of tin, had taken its place. I walked to the edge of a hill, which we crapped off of every morning. I looked at the spot where dung beetles had waited, quickly disposing of our contribution for the day. The beetles were gone. Memories of Lee returned—how he washed our clothes, his brown skin glistening in the sun as he worked in the field, the way he seemed to rear back as he surveyed the world around him with head held high and hands resting on his hips, and the ability he had to take all things in stride, no matter how bad they might have been. I loved the man. And he loved me as well. Any God without a place for a man like Lee in his kingdom can leave me out too.

We went to the fields, now grown up in *huisaches* except for a few acres planted in corn and beans. The lake was probably three-quarters full and stretched at least a mile from one end to the other. I'd love to go back and help them farm this land. A fertile field with water only a few turns of a valve away is a dream of any West Texas farmer. The men invited me to do so, telling me it would be allowed if the members of the *ejido* voted to allow it and they assured me they would.

But there are others—others who resent the intrusion of an American, others who become jealous when a community suddenly becomes successful, others interested only in subjugating and exploiting the labor force—who consider anyone giving them a legitimate fair wage a threat. I once went thinking I could change this and brought a great evil upon the community. Granted, what I did was illegal, but would it have been any different if it wasn't? I don't know. I suspect not.

The trip was good for me. I saw that things remain bad on both sides of the river. But I also saw a people determined to survive: a resilient, strong people, working together—unlike other places I see in my travels—a community, *en común,* a Mexican might say. I brought back a little piece of that community's spirit in my heart.

I passed through another community on this side of the river on my way home that is similar—the tiny town of Balmorhea—but here the story is even more tragic. An entire generation is gone, either in jail or dead. And the next generation is not far behind. Only now it's not just the boys who are involved but also the girls. When Crocket's name is mentioned it's not, "He died," but, "We lost Crocket." For me the town cannot be the same without him. The list of victims is long and the stories heartbreaking. The local librarian was almost in tears as she recounted their fate.

* * * * *

And then, the day after we return from Mexico, I go to court to watch my son, the flesh of my flesh and the blood of my blood, get sentenced. He's been in trouble again, not the kind of trouble some of these young people are in, but soon enough he could be.

The room is full of their young faces—beautiful young children of all races, but unlike those walking the street, these all wear orange jumpsuits. Chains encircle their waists, laced through handcuffs. Shackles restrict and bite into their legs. Anxious parents peer from the crowd. Muffled sobs escape their lips while armed guards stand ready, watching closely.

One blond-haired, blue-eyed boy in particular catches my eye. He seems distraught. His gaze moves from his mom to the judge and back again. I watch as one by one the kids go before the judge. When this boy's turn arrives, I learn that he has re-offended, multiple times. The judge shows no mercy, sentencing him to TYC, the prison for kids in Texas. Tears erupt from the boy's eyes. He shakes his head and tries to cover his face and his shame, but the handcuffs will not allow this. His mother has to be dragged from the courtroom as she cries uncontrollably. Tears of agony stream down her face. Groans rise from deep within and erupt into the room. The rest of us look on in horror, waiting our turn.

There are no winners in this war.

VICTIMS ALL: AN APPENDIX

The following is a partial list of those people who participated in this business with me and what I know of their fate:

Oscar Cabello—arrested and sent to prison in the United States for involvement with the Colombian-Mexican operation. He is now out, but his place in prison has been filled by his son, who was given fourteen years for marijuana possession.

Sergio Cabello—arrested and sent to prison in Mexico along with *eleven* others from Piedritas, all caught processing marijuana I grew.

Vicente Cabello—now a minister of the gospel.

Phil Ford—arrested and sent to prison in the United States. Now out, he works at a nursery, planting things and distributing flowers to the world.

Kenny Schulbach—arrested and sent to prison in the United States on unrelated drug charges.

James Risenhoover (a driver for Phil and me)—arrested and given probation.

Jeff Tyree—arrested and sent to prison on unrelated drug charges.

Steven Garcia—arrested and sent to prison on unrelated drug charges, multiple times.

Martín García—shot himself in the head rather than face arrest for a crime he committed.

Leroy Hernández—arrested and sent to prison on unrelated drug charges, multiple times.

Goo and Tato—prison, multiple times.

Crocket—married and sired a son. Killed in a car wreck between Balmorhea and Fort Davis, Texas, while drunk. A song by Johnny Cash comes to mind: *Call him drunken Ira Hayes, he won't answer you any more, not the whiskey drinking Indian, nor the Marine who went to war.* R.I.P. Crocket. You are not forgotten.

Amado Carrillo—killed on an operating table while undergoing plastic surgery on his face. Some think the body in the casket was not his. I'm not one of them.

Pablo Acosta—killed in a shoot-out in Santa Elena, Chihuahua.

Alejandro Cerna—arrested and sent to prison. He tried to make a deal with the government to testify against Lee Cross and me (and others, I'm sure) in exchange for leniency.

David McCasland—sent to prison for a very brief time (less than a year).

Coylle Lee Cross—arrested and sent to prison where it was discovered he had lung cancer. Was in remission and doing fine when I last saw him. Present whereabouts unknown.

Jeff Aylesworth—unknown.

James Robertson—unknown.

Arnold Kersh—killed in his own home when he was twenty-four.

Dick Graham—shot and killed in Sheffield, Texas, while trying to flee the country.

Tom Graham—shot in the face and blinded, given a thirty-five year sentence in TDC.

David Regela—now "retired."

Michael Palmer (pilot of the DC-6)—made money from all sides of the issue: the traffickers, the DEA and the CIA. Last I heard he was running a transport company out of Florida by the name of Vortex. Never served time in the U.S. for his illegal activities.

Buzzard—shot between the eyes with a .44 magnum.

Glenda Cross—spent several weeks in the custody of the Mexican authorities. She talked.

Cheryl (my ex-wife)—raised our children the best she knew how. But how do you support a family and be there at the same time?

Macario Villarreal—turned out to be the local snitch that set us up, first with the army, and then later with Acosta. He sits in a Mexican prison. He's fortunate that Oscar is a good man.

Beto—poor but alive and well in the northern mountains of Mexico.

Ann Steinmetz—still tirelessly defending those without money with all she has to offer.

The people who bought and used the drugs we and others like us supplied—in essence financing us—can be found in just about any walk of life. One is now President, another an ex-President, but then, he didn't inhale.

* * * * *

I received a total of fifteen years for my crimes. Under current law, it would have been much more, perhaps in the neighborhood of twenty years, and I would not be eligible for parole. My children grew up without a father and bear the scars even today.

I think it fair to say that none of us emerged from this business unscathed.

I think it is also fair to say that we all—the smugglers, the dealers and the whores—have been replaced, and that a similar or worse fate awaits the present-day crowd involved in this business. And then they will be replaced.

THERE is no foreseeable end to this story,
not from my vantage point.